Introduction
to
Scientific
Geographic
Research

Haring & Lounsbury

Introduction to Scientific Geographic Research

Introduction to Scientific Geographic Research

L. Lloyd Haring

John F. Lounsbury

Arizona State University

WM. C. BROWN COMPANY PUBLISHERS
Dubuque, Iowa

Copyright © 1971 by
Wm. C. Brown Company Publishers

Library of Congress Catalog Card Number: 74-145773

ISBN 0–697–05281–8
Second Printing, 1972

Printed in the United States of America

contents

list of tables

list of figures

list of appendixes

foreword to the student

This book is addressed to undergraduate majors in geography and to beginning graduate students who are about to undertake, for the first time, some basic research in their major field of specialization. This does not imply that, prior to this occasion, you have not read widely, observed events at first-hand, or produced scholarly and worthwhile papers based on reliable sources of information. It does assume, however, that these papers did not require you to be as reflective, logical, or scientifically directed as does the assignment you are about to undertake. Now as a true research scholar in geography, although a beginning one, you have been asked to use the analytic mode of thought, commonly referred to as the "scientific method" to solve a particular problem in which, hopefully, you are much interested. This book sets forth the steps that you will need to follow if your thought patterns are to become tutored rather than untutored, disciplined rather than undisciplined.

As you develop your ability to use the analytic mode of thought, keep in mind that it is *one* approach to knowledge, but certainly not the only one. Not all research or inquiry need be scientific. There is, for example, the artist's approach to understanding, a purely subjective approach, and there is what might be called the "common-sense" approach. In the past, much geographic research was not undertaken objectively as required in the scientific method. Rather, events were observed and interpreted as comprehended by the observer, that is subjectively or artistically. Rarely did two geographers, confronted with the same interests, collect comparable data or arrive at the same conclusions. This does not indicate that one was "wrong" and the other "right." It simply indicates that their procedures and findings were somewhat personal; their interpretations intuitively derived or based on subjective reactions to observed events. As geographers, they did not proceed in their investigations in ways that others could follow to arrive at similar conclusions.

Recently, geographers have moved away from using non-scientific modes of inquiry, and have begun to investigate questions and problems more objectively. Their work now meets the standard of replicability, and their aim is to have their work contribute to the development of a logically consistent, conceptually-oriented frame of reference. Often this frame of reference is expressed in such forms as theories or models. As a result, geography is emerging as a more abstract, theoretically-oriented science involving the analytic method of inquiry.

The widespread adoption of a more scientific geography during the 1960s was made possible by the general spread and growth in the use of quantitative techniques, both statistical and mathematical. Important statistical techniques for analyzing spatial processes are presented in this book.

As a student, you should be forewarned that there is nothing in this book that can be called new to those who have earned the doctorate in geography within recent years, or who are carrying on basic research in geography using the scientific method. The steps outlined are followed conscientiously by mature scholars who have become recognized for their contributions to scientific research in the discipline. It behooves the beginning student, therefore, to follow the steps in geographic research which the authors of this book have described. Their discussion of these steps presents in very practical terms a systematic and basic plan for research in geography.

Clyde F. Kohn, Chairman
Department of Geography
The University of Iowa

preface

It is likely that college students majoring in geography will first be exposed to geographic research methods at the advanced undergraduate level or at the beginning graduate student level. Many geography departments offer either formal courses concerned with introductory research methodologies or informal courses such as seminars, tutorials, or guided study courses. At the present, there is little published material designed specifically to serve as a guide for students involved in their first actual research project. Often the student is frustrated in his attempt to make order out of chaos. He finds it difficult to define precisely his research problem and associated hypotheses, to collect meaningful data, to employ the proper techniques of analysis, and to come to logical conclusions.

Modern geography is a discipline with many recognized and well-established subfields. Many of these subfields have developed special research techniques utilizing highly sophisticated data-collecting and data-processing equipment. It is not the intention of the authors to provide a book to include the advanced research techniques and tools applicable to all the subfields within the discipline. Rather, the major purpose of this book is to serve as a guide to enable the student to develop an orderly, scientific outlook and to provide a framework of reference to direct the student's enthusiasm and efforts along productive tracks.

The book is designed for students who have little or no previous research experience. It may be used in any course concerned with introductory research methods. It also may serve as a useful supplementary text in field technique courses, as well as an aid in preparing term papers or projects in a variety of other geography courses. The book has been designed purposely to be general rather than specific, broad rather than restricted, and flexible rather than rigid. In this way, the instructor can select what is useful to his specific needs, as well as add substantive material without difficulty.

The authors have taught introductory research methods for many years at several colleges and universities, at both the undergraduate and graduate levels. The authors received their graduate training at different universities, and each has his own areas of specialization. It is hoped that this diversity in the authors' training, viewpoints, and professional specialties is reflected in the book's broad approach, pertinent to the overall discipline of geography, rather than in emphasis on specific subfields.

L. L. H.
J. F. L.

the nature
of scientific research

It is an accepted part of twentieth century thought that science and its research methods are of social value. This does not necessarily mean that science is the only path to human wisdom. But it is evident that the path to human satisfaction (and by this is meant satisfying the desires of modern society) is paved with the discoveries and recorded contributions of science. Whether the desire is to go to the moon, to cure the sick, to build a better automobile, or to take the salt out of ocean water, man turns to science for its solution. It is for this reason that the social and physical sciences are important and respected segments of higher learning today.

The Framework of Science

Science seeks the truth in an objective, rational manner through a process of controlled inquiry. The guiding impulse of science is curiosity, and the beginning point is a question for which an answer is sought. Throughout this search, all terms used are defined precisely, and all procedures are described carefully. All processes of deductive or inductive reasoning are clearly demonstrable, and each tool or technique employed is explained. Scientific conclusions are never expected to be accepted on faith or solely on the authority of the researcher. Clearly, research in science is not for everyone. It is only for those who are willing to demonstrate the validity of each step, from the posing of the question to its eventual solution. It is for those who not only want to know, but who will not be satisfied with an answer unless the process and reasoning behind it can be verified by reason, analysis, and rigorous testing.

The "scientific method" is commonly credited to Sir Francis Bacon (1561-1626) who set forth its tenets in his work *Novum Organum* (or *The New Method*).[1] Prior to this time, answers to questions had been accepted primarily by unquestioned belief in authority, either based on faith or the conjecture of logic. Bacon expressed the opinion that these methods of problem solving were often wrong. He espoused a process which began with the problem for which a solution or "hypothesis" would be suggested. This hypothesis, assumption, or educated guess then guided the research through observation, analysis, synthesis, and finally, to the conclusion. The conclusion, then, would come only after the hypothesis was thoroughly tested. On the basis of the evidence gathered, it was accepted or rejected as an answer to the original question. During this process, the scientist remained objective, suspended judgment until all evidence was gathered, and recorded his data carefully and precisely.

The scientist today follows procedures very much like those set forth by Bacon. His answers are always objective, verifiable, and impartial, and whenever possible, they have mathematical precision. He is not content to deal in vague terms, such as "more" or "less," but seeks to discover exactly how much more or how much less. He weighs, measures, and calculates systematically, each step logically following the one before. The final answer is based on the preponderance of evidence carefully analyzed and is as correct as the facts available at a given time will allow.

Methods of Research

The scientific researcher today selects the procedure to fit the problem with which he is concerned. In general, three major "methods" of procedure are recognized: the *experimental*, the *normative*, and the *historical*.

The first method is used mainly by the physical scientists, such as chemists and physicists, who are able to control the factors involved in their research and thereby control the processes which result. By rigorous control of all variables employed in the experiment, the physical scientists are able to describe, analyze, and predict with a high degree of precision. Their work has become the epitomy of science for many people.

1. James Spedding, Robert Leslie Ellis, and Douglas Denon Heath, eds., *The Works of Francis Bacon* (London: Longmans & Co., 1879), vol. 1, pp. 70-117.

In those areas of research where variable control is not possible or feasible, the normative method is a widely used method of investigation. The essence of this method is to observe events and evaluate the observed processes with a view of establishing constant relationships or norms. In this method, description and analysis are as important as they are in the experimental method. Where relationships are discovered, prediction becomes possible. Frequently, geographic norms take the form of a dependent variable "Y" varying proportionately with an independent variable "X." From such a relationship, laws based upon these norms may be formulated, and their predictable variation may be stated.

A third method of research is one in which the researcher neither controls nor evaluates the variables observed. Here the major objective is to record observations accurately. The historical method relies heavily on source detection, evaluation, and analysis of findings in determining solutions or conclusions to the research problem. An excellent example of this method is found in medical research where a relationship between death and stimulating social events has been established. Between 1875 and 1915, the death rate in Budapest, Hungary, dropped markedly before the observance of Yom Kippur. From this and corroborating information, the evidence suggests that dying is a form of social behavior.[2]

While widely used by historians, this method is often combined with the normative method in historical geography studies. By this means, the researcher may describe historical events and establish past spatial distributions and areal associations.

Geography as a Research Discipline

Man is curious about areas and regions which are different and removed from his immediate environment. Mankind, however, does not possess an inborn instinct or innate knowledge of areas. This knowledge is acquired only through education, formal and informal, over a long period of time. The field of geography is primarily concerned with the acquisition of this kind of knowledge within a scientific and highly structured framework. It is the major discipline that is concerned with the identification, analysis, and interpretation of spatial distributions of phenomena and their areal associations as they occur on the surface of the earth.

2. David Phillips, "The Vital Buoyancy of Optimism," *Time*, September 5, 1969, pp. 58-60.

The field of geography as a whole does not specialize in one particular set of phenomena.[3] While most other sciences are focusing their studies on a process, an event, or an object, geography is characterized by its method, in addition to its substantive material. Like science itself, the discipline focuses on method. All phenomena that occupy space are grist for geographic analysis. The goal of all science is to describe, analyze, and finally, to predict. Geography does these things and is capable of doing them very well. Geography is a science because its research techniques utilize the scientific method and produce results similar to other scientific disciplines.

Geography is a vigorous science concentrating on the concept of space and spatial relations which, along with time and composition of matter comprise the three major parameters of concern for all of science:

> Of the three great parameters of concern to scientists, space, time, and composition of matter, geography is concerned with two. Geography treats the man-environment system primarily from the point of view of space in time. It seeks to explain how the subsystems of the physical environment are organized on the earth's surface, and how man distributes himself over the earth in his space relation to physical features and to other men. Geography's organizing concept, for which "spatial distributions and space relations" are a verbal shorthand, is a tri-scalar space. The scales comprise extent, density, and succession. Geography's theoretical framework is developed from this basic concept.[4]

There is reason to believe that geography is perhaps the most logical discipline for understanding the systems of life, in their totality, as geographic research investigates and tests hypotheses advanced to explain problems of spatial associations. It is essentially concerned with place, location, territory, distance, space, and their interactions. When the methods of science are applied to these problems, the results are worthy of interest to the entire scientific community.

There will always be workers in any scientific discipline who wish that their areas of specialization were not a science,

3. Jan Broek, *Geography: Its Scope and Spirit* (Columbus, Ohio: Charles E. Merrill Publishing Co., 1965), p. 5.
4. Ad Hoc Committee on Geography, Earth Sciences Division, *The Science of Geography* (Washington, D.C.: National Academy of Sciences-National Research Council, 1965), p. 1.

for science is a stern taskmaster. For those who wish to rely on opinions, who want instant answers, or who prefer conjecture or faith as a method of gaining answers to problems, science is not the suitable approach. The goal of scientific research is not a collection of interesting observations about some place, event, or process. Such studies may be artistic, interesting, fun to do, and even useful; but they do not fall within the realm of scientific geographic research.

Whether it be a social or a physical study, geographic research is the application of the scientific method toward solving spatial questions. As such, it must have a problem, an hypothesis, accurate data, a presentation of findings, and a conclusion. It may describe or analyze or predict; or all three purposes may be served in one study. But it must follow the procedures of science if it is to be considered a scientific research project.

Steps in Geographic Research

Successful research in geography consists essentially of completing a series of specific tasks or steps. It is important that they be done in a proper order to avoid confusion; for as each task is completed, it determines the specific nature of other steps immediately following. The amount of time and effort expended on any given step varies depending upon the nature of the research problem. In general, these steps in the order that they should be worked upon may be defined as follows:

(1) *Formulation of the research problem*, or the asking of a previously unanswered question in exact terms. This also includes the precise determination of the areal extent of the matrix within which the research work will be conducted. A research problem may be concerned with a micro-area, such as a city block; or it may be concerned with a macro-area as large as a continent.

(2) *Definition of hypotheses*, or formulating a theory, assumption, or a set of assumptions which are yet unproven but which are accepted tentatively as a basis for investigation.

(3) *Determination of the type of data to be collected* pertinent to the research problem. The specific nature of the research problem and the size and areal extent of the matrix will determine the type of data that must be obtained, the way the data will be classified, how the data will be collected, and whether or not the matrix will be surveyed in its entirety or sampling procedures will be employed.

(4)   *Collection of data*, which involves the analysis of published materials, the use of field techniques, or perhaps the deployment of data-collecting instruments.

(5)   *Analyzing and processing the collected data*, which involves the selection of appropriate cartographic and statistical methods of analysis. Further, this final step includes stating the conclusions and determining if the proposed hypotheses are confirmed or denied.

In the chapters to follow, these procedures will be examined and discussed in detail.

chapter 2

defining
geographic problems

The first step in any research plan is the careful selection and identification of the problem. Problems usually arise as the result of a feeling of concern or the need for more precise knowledge. Wherever the student notes a lack of knowledge in the corpus of geography, a potential problem exists. A proper research problem is a situation of concern in geography that is described, bounded, and focused in order that concentrated study may be applied to it. For example, while studying urban distributions, a geographer might notice a lack of locational information concerning cities between 100,000 and 200,000 population. If he feels that additional information is needed before an understanding of urban distributions can be acquired, he has experienced a *problem*. Since problems are actually unsolved questions, one way to define the specific research problem is to ask a question such as "What are the relationships between urban sprawl and internal population mobility in cities of 100,000 to 200,000 population?" Assuming that the results of the study are meaningful, the research has contributed some knowledge that will be of value in the understanding of the overall question of the areal dynamics of medium-sized cities.

The number of "good" research problems in geography is infinite. Many times, such problems are defined by scholars in books, in dissertations, or in professional papers. In fact, it is a somewhat standard practice to conclude a study with suggestions for further research. The serious student of geography will analyze published materials concerning his subject, as well as materials developed by related disciplines. From the subject matter or the principles presented in the literature, he may identify gaps which appear to justify additional research. There is no better way to formulate a problem than for the researcher himself to sense its need and state it in his own words. In doing this, the first step is to select specific phenomena whose locational presence in an area is of concern, then proceed to explain their presence and spatial arrangement. The phenomena to be studied may be physical or cultural, concrete or abstract, objects or ideas. They may be house types, drainage patterns, votes, or religious concepts. Any material or nonmaterial thing which can be identified, classified, and located is proper subject matter for geographic study. The geographic problem contains the elements of what, where, and why— a distribution of phenomena (what) whose locational or spatial characteristics (where) are to be explained by their association, or lack of association, with other phenomena (why).

The mere existence of an unanswered question does not necessarily assure the basis of a suitable research problem. There may be inherent difficulties concerning a problem, difficulties which will raise serious doubts as to its research feasibility. Is it of interest to the researcher? Probably it will be if it has been discovered and developed by his own efforts. But if it is not interesting to the individual researcher, it is doubtful that the research should be undertaken. This is especially true in theses and dissertations where considerable time and effort will be required. The quality of work will probably suffer because of the tedium of the research work; and the lack of interest may be related to other factors which will seriously affect the product.

One possibility which could detract from the interest of a problem is its apparent lack of value for a given individual. If the student does not have some curiosity about a problem, that problem probably does not have sufficient motivating value for the student to expend the sustained research effort necessary for its solution. The main object of a research project is for the apprentice research worker to learn to do research and to demonstrate this acquired skill. The results of the research need not be of practical value, nor need they make a significant contribution to the discipline. They must, however, be of some importance to the student himself, even if only to satisfy his intellectual curiosity.

The beginning research student runs a special hazard in this respect. Quite often, his uncertainty makes him especially susceptible to suggestion. While a conscientious advisor is careful not to pressure the advisee into an unsuitable topic, such a situation can occur. The student is well advised to have a topic of interest to him in mind when he discusses the matter with an established researcher. In this regard, the late Professor Good describes a situation which is not uncommon. The beginning research student came into Professor Good's office, and this was the ensuing confrontation:

> "I've got to write a Master's thesis," says he, "and I'd like to talk to you about a topic." The statement ends with a slight upward inflection as if, in spite of its grammatical form, a sort of question were implied. After an awkward pause Mr. Blank (the student) repeats that he would like to talk about a thesis topic. Whereupon the Editor (and Professor) suggests that he go ahead and do so.

> It transpires, however, that the Editor-Professor has misconceived Mr. Blank's meaning. He has no topic to talk about. In fact, instead of coming with a topic, he has come to get one. He looks so expectant, too; purely, as one might say, in a receptive mood. . . . He gives the impression of having just learned about this thesis business, and of being entirely open-minded on the subject.[1]

This account, while containing an element of humor, occurs too frequently to be completely humorous. If the research project is to be a pleasant exercise, as it should be, the problem should be the writer's choice and should reflect his personal interest. The student should profit from his advisor's suggestions, but he should not depend on him to define his problem or to outline the study.

After it is decided that a problem exists and that a solution to the problem will be of value to him, the research student should determine whether a solution is possible or even probable. For instance, one might think of very interesting problems, such as determining the effect of the lost continent of Atlantis on past climatic conditions in Europe. However, the possibility of testing any hypothesis set forth as a solution for this problem is so remote as to make it unacceptable for research work. Serious reservations should be entertained if the problem is of such a nature that a solution is not likely, or if the qualifications of the researcher are not adequate to solve the problem. In some cases, the lack of qualifications may be corrected by the acquisition of new skills or by mastery of additional research tools. Even experienced scholars find it beneficial at times to return to the

1. H. G. Good, "The Editor Turns Professor," *Education Research Bulletin*, VI (September 14, 1927), pp. 252-53.

classroom in order to learn skills necessary for the solving of new problems.

A common weakness with many interesting problems is that sufficient existing data is lacking, or new data cannot be obtained for a successful solution. This is the weakness of the hypothetical problem concerning Atlantis; and many contemporary problems have the same inherent weakness. While it may be possible to obtain the needed data by extensive field study, the beginning student should be aware of the magnitude of the task. He should carefully inventory his resources—time, money, knowledge—and if the problem cannot be solved within his limits, he must either redefine and reduce the scope of his problem or implement his resources.

In the last analysis, a satisfactory research problem in geography is one which is of interest to the researcher, one in which sufficient data concerning it may be obtained and which focuses on areal associations and spatial relationships.

In every stage of problem development and articulation, it is imperative that the student read widely concerning the subject. Any literature in geography or related fields which touches upon the problem should be reviewed (see Appendix A). The reasons for this are obvious, as the researcher must know what type of data is available, or may be obtained, and whether or not the problem has already been investigated. A satisfactory solution to a problem may present new data; it may develop a new methodology; or it may bring existing data together in a new way to make it more meaningful. Whatever the primary focus of the problem, the researcher will not wish to invest his valuable time solving one that has been solved previously. Nor would he desire to fail in a solution because the problem could not be solved. Squaring a circle or inventing the wheel may seem to present problems, but one is impossible and the other unnecessary!

Determining the Matrix

An important aspect of formulating the research problem is determining precisely the areal extent of the research area or matrix. There is no standard or generally accepted size of a matrix, as significant research may be accomplished in very small or very large areas. However, the size of the matrix will bear directly on the types of phenomena that can be studied, as well as on the scale, detail, and classification of the data to be collected.

For example, a student researcher might become interested in changing land use patterns in the United States and pose the question "What is the relationship between types of land use and population densities in the United States?" If the problem is so stated, it would be in order to illustrate the spatial distributions of land use and population. It is possible to construct such illustrations for the entire country but only by using very general categories of land use, such as urban land, agricultural land, forest land, grassland, and the like, and by using broad categories of population densities. However, at this scale, the resulting maps would be so highly generalized that no definite conclusions could be drawn for specific areas.

Consider the situation in which the student is still concerned with the same general problem but reduces the study area to a county in size. It is now possible to classify land use data in more explicit terms; and what was urban land may now be subdivided into residential, commercial, industrial, and so forth. What was agricultural land may be divided into cropped land (even the type of crops), pasture land, fallow land, and the like. In the same manner, the categories of population densities may be made more detailed. This information shown spatially results in maps that are more refined and accurate than are those for the country as a whole, and the relationships become clearer.

From the results of this hypothetical study, the student may want to pursue the matter further, desiring to know how commercial land use in the central business district of the city and commercial land use in outlying shopping centers are related to population densities. The matrix may now be reduced to a few square blocks in size. This makes it possible to obtain the actual square footage of each commercial use, as well as the exact population, block by block. He may find that some commercial uses are always located close to dense populations, while some types of commercial uses may be relatively far removed from residential areas.

The matrix is the framework within which the research is conducted and which determines the detail of the data collected. Usually, the smaller the matrix, the more specific will be the data; and the larger the matrix, the more general will be the data and the likelihood that sampling procedures will be employed.

In determining the matrix, the delimitation criteria used and the rationale for selecting this particular research area over all others should be stated clearly. For example, if a research problem is concerned with some aspect of wheat farming in the Northern Great Plains, the first questions that will be asked are, where are the Northern Great Plains, and what delimits them? why not study the entire Great Plains area? or, why not study only the southeastern quadrant of the Northern Great Plains. Often, the matrix of study may be justified on the basis that superficial evidence indicates that a given phenomenon or activity appears to be concentrated in the proposed area more so than in others, or that there exists a widespread feeling of concern about a particular situation in the proposed area, or perhaps that there is a need for new information about the proposed area.

The boundaries of the research area may often coincide with a physical feature, such as a valley or a river basin, or a political division or other cultural boundary. The matrix need not always be a contiguous area, as comparing two or more areas is not uncommon. Also, the definition of the boundaries of the matrix may be influenced because essential existing data are available in one area but not in another. For example, one county may have accumulated information which is highly significant to the research problem, but this information is lacking in adjoining counties or areas.

Stating the Problem

Once a problem and its matrix are judged suitable for research, the problem should be stated carefully. At this point, it is essential and natural that a positive, confident frame of mind exist. The student has determined that the problem is pertinent, interesting, and that it can in all probability be solved within the limits of his resources. This is the problem, and he is now ready to state in exact terms what it is he intends to do. He will need this statement for his own purposes and so that others will know precisely what he is going to do and what he is not going to do. The success or failure of the research project may depend upon the proper statement of the problem.

There is no one way in which a problem must be presented, but its statement should always be clear and concise. There should be no doubt as to the need for this research nor as to the subfield of the discipline which will profit from its solution. A question is one good way to pose a problem. Frequently, when one wants an answer, a question is asked. Such questions are to the point and afford pegs on which answers may be hung. If the problem is presented as a statement, it can be changed in form to be a question. For example, the geographer may present a problem concerning the relationship between cash grain farming and flat land as a statement, or he might pose the question, "What are the relationships between cash grain farming and low slope land in the American Corn Belt?" Since any problem is by its nature an unanswered question, this latter form is an accepted scientific method of posing it.

There are some important aspects to be noted in the example just given. In the first place, a complete statement of problem must define the terms which it uses. Other persons may have no way of knowing what is meant by cash grain farming, low slope land, or the American Corn Belt. Until these things are known, and in terms which can be tested for possible solutions, it cannot be determined if a problem exists. Thus, additional questions must be asked, such as, what is low slope land? what are the parameters of slope? what is cash grain farming? what is the areal extent of the American Corn Belt? and, what tests will be used to determine if a relationship exists? In answer, a cash grain farm might be defined as one on which 50 percent or more of all farm income is from the sale of grain. The American Corn Belt might be defined as the area within which 20 percent of the cultivated land is devoted to corn production any given year. Low slope land might be defined as having slopes of 1° to 5°.

This procedure of defining terms is referred to as "operationalizing" the problem. Until the terms used are described in such a way that all can understand what is meant, there is no way in which a testing operation can be performed on any hypothesized relationship. In the following section, the formulation of suitable hypotheses will be discussed. For now, it is sufficient to keep in mind that the use of words or terms not susceptible to an operational definition usually remain nebulous, if not meaningless. "If no operation can be performed, it is highly doubtful that two human minds can get close enough to the subject to discuss it intelligently."[2] Now that it is known what is meant by low slope land, cash grain farming, and the American Corn Belt, the researcher can proceed to establish the type and degree of relationship existing between the two variables within the defined matrix.

Many problems in geography are questions concerning the spatial relationship of two or more variables. Normally, the question to be answered must be stated in such a way that the tentative answer, the hypothesis, may be tested to determine if it will

2. Stuart Chase, *Guides to Straight Thinking* (New York: Harper & Brothers, 1956), pp. 28-29.

be accepted or rejected—or with what reservations it may be accepted. The formulation of the problem, then, is the first step in research; and it is wedded in its inception to the hypothesis. The scientific research problem exists in order that a scientific research solution may be sought. If the research problem is clearly stated and the research area well defined, the remaining steps may be time-consuming; but the researcher can rest assured that his subsequent efforts will be organized and that they will be expended along the most productive lines.

Nature of the Hypothesis

It is possible that some research problems may not have a formally stated hypothesis. If a research study is concerned with a problem for which little or no information exists, it may not be possible to formulate any reasonable assumptions or hypotheses. The formulation of hypotheses implies some knowledge of the problem and the research area. For example, if research is proposed to determine the spatial distribution of permafrost depths in Greenland, or the cultural traits of primitive peoples in the Amazon Basin, there may not be sufficient existing data upon which to base meaningful hypotheses. The answers to basic questions such as these would contribute, in themselves, to the overall pool of knowledge and would be accepted as worthwhile research.

It is not likely that the beginning student researcher will have the qualifications or the opportunity to become engaged in the type of research problem just described, one for which almost no information would be available. Ample information of some kind, either in the realm of general knowledge or in related research work, normally will be of a sufficient quantity to enable the researcher to formulate one or more hypotheses.

The statement of the problem is the identification of a felt need to know. The hypothesis is a reasonable way to meet that need. It is a proposition which is assumed to offer a possible and reasonable solution to the problem. Together, the problem and hypothesis guide the research investigation.

Certain characteristics may be observed in a good hypothesis. First, it is capable of being expressed as a question. For example, the hypothesis that temperature change is associated with elevation may be expressed as the question, "Is temperature change associated with elevation?" Second, the hypothesis may be stated in a negative way: "Temperature change is not associated with elevation." This type of hypothesis statement is a *null* hypothesis and is useful in certain problems. Third, the hypothesis should be capable of being answered with a "yes," a "no," or a "maybe." In formal terms, it is capable of being accepted, rejected, or not rejected. The difference between these hypotheses and everyday questions all people ask themselves about observed phenomena is one of degree, and not of kind. As stated in Chapter 1, the scientific method is no more than sound reflective thought applied to a specific question.

In research work, the hypothesis is a statement which might be called an educated guess, an informed hunch, an assumption, a suggestion, a supposition, or a conjecture. Whatever it is called, its usefulness depends upon how well it condenses an array of facts into a statement that can be investigated and tested. It is this investigating and testing that elevates it to the status of an hypothesis. The better the researcher understands the facts related to the problem, the more he knows about the research area. And the more educated his "guesses" or hunches tend to be, the more likely he is to identify correct solutions and to direct his efforts into fruitful lines of research.

If one were attempting to chart the shortest route by automobile to New York City from Los Angeles, he would be faced with a large number of possible alternatives. Many could be eliminated as being too improbable to consider (or "test," in research language); but a myriad of routes and subroutes would still remain. Perhaps someone who has just returned to Los Angeles from New York claims that the shortest way to New York is through Denver. This may be a reasonable supposition, and untrained people accept such informed guesses and suppositions as truth. However, until it is tested against other probable ways, the Denver route cannot be accepted as the shortest way. After it is measured against a possible shorter way, it may then be rejected or accepted as the shorter of the two routes. But it cannot be accepted as the shortest route on the basis of comparison with only one other route. As other reasonable possibilities are eliminated, there will be a point at which the hypothesis may legitimately be accepted as correct. The reason for the hypothesis becomes more apparent when one attempts to solve a problem without one.

The hypothesis is not always stated as a clear and separate corollary of the research problem. When it is not, the detection of the precise purpose of the research may be more difficult. For example, a problem may be stated in a manner that includes the hypothesis: "The purpose of this paper is to determine if Republican voters in New England tend to be located in rural areas." The problem here would be accounting for the locational pattern of the Republican voters in New England, and the hypothesis is that this distribution is associated with "rural" conditions (precisely defined). A slight modification sometimes encountered is the making of the statement in the form of an hypothesis: "It is hypothesized that rural conditions account for the locational pattern of Republican voters in New England." In both cases, it would seem that a more easily understood method of informing the reader of the exact nature of the problem and the hypothesis would be to state the problem clearly as a question, and then, in a separate statement, advance the hypothesized answer.

Determination of Hypotheses

Usually, there are two ways of arriving at tenable hypotheses in geographic studies. One is to consider the distribution of the phenomena which the researcher wishes to explain and to seek reasonable explanations in the laws that control that phenomena. The other is to consider the location and density of the phenomena and to compare this pattern with the pattern of other supposedly related data. Each of these methods warrants individual consideration.

In geographic research, the starting point is the distribution of some phenomena which a researcher wishes to analyze in terms of where they are and why. This is called a dependent variable since its locational pattern depends on some other variables (phenomena) which effect it. For example, in explaining the location of steel mills, two related variables might logically be iron and coal sources. The first method, then, of arriving at logical hypotheses is to turn to the laws which govern the areal location of the dependent phenomena and to cognate how much of the activity should be expected at a certain place. By knowing the proportions of iron, coal, and other materials in a ton of steel, and through a process of evaluating this information in relation to transportation costs and markets, a reasonable hypothesis could be formulated. It would probably take the form of "At point X, a steel mill would be expected." In this case, the hypothesis could be tested by observing how closely the theoretical location corresponds with actual location.

It is obvious that such hypotheses require considerable knowledge or information before an acceptable solution to the problem of the location of a dependent variable can be expected. Because of the complicated nature of this information and the need for precision, such hypotheses are often stated mathematically with the use of numbers and symbols. Such statements are called *mathematical models*. Just as a table model of a house may express accurately the dimensions of someone's home, so, too, a small conceptual model which is stated mathematically may express a real life situation. For example, the amount of travel expected between City 1 and City 2 might be predicted by considering the populations of the two cities and the distance between them. This is done by the "interaction model" which is stated as $i = \dfrac{P_1\ P_2}{d}$ where "P" is the population of City 1 and City 2, and "d" is the distance between them. The interaction model, the simplest form expressed here, has been tested and refined, until today its many forms have a certain degree of accuracy in explaining actual transportation patterns.

Geographic hypotheses are often suggested as a result of the geographer's comparing the distributional pattern of a dependent variable with other known distributions. In this method, the dependent variable is usually mapped, and then this map is visually compared with other maps. On the basis of this comparison, an "educated guess" is made concerning the association between two of the maps. This guess is then studied in order to determine if an association does indeed exist, and if so, to what degree. As a rule, the locational arrangement of a variable under consideration is affected by many other variables acting upon it. The researcher must assemble a sufficient number of these independent variables that might explain the pattern of a phenomenon he proposes to study. For example, if the problem is "Why is the distribution of crime in Century City arranged in this particular pattern?", the possible hypotheses may include the distribution of: (1) schools, (2) income, (3) broken homes, (4) poor housing, (5) ethnic groups, and an infinite number of other variables. The researcher cannot hope to explain 100 percent of the distribution of such crime. Instead, he must attempt to select the significant variables effecting its location and settle for the degree of explanation desired in his problem design.

This method of hypothesis determination was used in the research on yellow fever in Panama. Maps of various possible related phenomena were compared with a map showing the areal extent of yellow fever until a likeness was detected between the fever map and a map of the distribution of the anopheles mosquito. A relationship was hypothesized, tested, and accepted. After elimination of the mosquito, yellow fever was brought under control, and the completion of the Panama Canal thus became possible. A more recent, but similar, situation occurred in a study of jaw cancer in Africa. In this case, a chance remark that the cancer was not found in southern Africa led to the problem: "Why does the tumor occur on the northern side of a given line across Africa and not on the other side?" Comparative map analysis detected an association between the line of cancer and that of rainfall, temperature, and eventually, a virus-carrying insect. Thus developed one of the major discoveries in cancer research in our lifetime.[3]

As the body of geographic literature develops, more laws governing geographic relationships may be expected. From such relationships will come the source of future hypotheses. However, for the foreseeable future, many of the hypothetical solutions will be obtained from maps, ideas gathered from reading and observations, commonsense deductions, and similar sources. These hypotheses will be tested by measuring the extent to which they express actual locational patterns. The tested and accepted hypotheses, after being subjected to extensive reexamination to confirm their validity, will become the laws of geography. Eventually, researchers will be able to turn to these laws for new hypotheses. With each tested and confirmed hypothesis, geographers will become capable of describing and analyzing the locational pattern of a specific phenomenon and of predicting its spatial relationship with other areally associated phenomena.

3. Bernard Glemser, "The Great Tumor Safari," *Today's Health*, XLVI (September 1968), p. 44 ff.

chapter 3

formulation of
the research design

After the student has defined his problem, matrix, and related hypotheses, he should take a reflective look at his entire project and then design an operation plan. This "work plan" might be compared to the blueprint that a construction engineer requires before beginning a new structure. It controls the phases of inquiry so that procedure decisions are made before a situation arises. The research operation plan includes not only the blueprint of the project to be undertaken, but also the materials needed, the tools required, the cost involved, and the time schedule of anticipated progress.

There are many reasons for making such a plan. In situations where an advisor, a committee, or a supervising professor are involved, presentation of the plan may be required before the proposed project is approved. If outside funds are to be sought, the funding agency will request such a plan well in advance of the proposed time schedule. Perhaps certain sources of desirable information may be denied to the researcher unless he can demonstrate the purpose of the research and the manner in which the information will be used.

The most important reason for the work plan is that it provides the framework to allow the researcher to organize his time and resources. Only the most hardy (and foolish) person would embark on an important task without knowing his goal, how he planned to reach it, and at what time he should be there. Houses have been built without a blueprint, but no modern, competent builder would attempt such a project. Nor would he proceed without deciding what parts were to be built first and at what approximate time each stage should be completed. In the building industry, it is common practice to post a bond stipulating the date on which a contracted structure will be completed. Financial ruin would rapidly overtake a contractor who worked without a blueprint and a timetable. Inefficient use of time and efforts is the result of research without a work plan.

The most important parts of a work plan are an orderly arrangement of the steps involved, a time schedule, and a list of equipment necessary for the completion of the steps involved. Also, the finished work plan may be presented to others who are involved or interested in the study, and it becomes a guide for the entire project. As the study progresses, specific parts of the work plan may be expanded, altered, or eliminated as new information becomes available.

The Title

The beginning of the work plan is the selection of the title of the research study. Although this may be a tentative title, subject to modification after the paper is complete, the most appropriate title for the proposed research should be selected at this time. It should be short, but not so brief as to obscure the field of study and the essence of the research. The title should indicate the subfield of geography concerned, the research matrix, the time period, and the purpose of the study. A title such as "The Relation of Seasonal Crop Harvesting and Migrant Labor in California: 1940-1960" fulfills the requirements of indicating space, time, subfield, and purpose. It also tells others at a glance the major variables to be studied.

Purpose and Problem

After the title, the work plan should contain a brief, but pertinent, description of the area of geography in which the proposed study falls. This description should include the need for the information anticipated in the study and the specific gaps in the literature which will be filled. It may seem difficult to envision results before the study is begun, but the formation of tentative conclusions is an important, and often a most neglected, step in research planning.

In this part of the proposal, major terms to be used are defined in such a way that other persons will know what the terms mean and how they will be used. If the study is of village population change, the reader must know the size of a village and how much population increase or decrease is necessary before a "change" is indicated.

Whenever possible, the anticipated direction of study should be stated simply. If, for example, the problem is to determine the effect of "X on Y," this should be stated. The researcher may be interested in the effect of rainfall (X) on corn yields (Y), or income (X) on Republican votes (Y), or transporting costs (X) on gasoline prices (Y), or any other phenomena that may be identified and measured. Let us assume that a study is focusing on the variation of the maximum temperature at a number of locations in the researcher's home state on a particular day. Further, let us assume that the researcher has hypothesized elevation as the factor associated with temperature variation. He can clearly and simply inform others of the proposed problem in the following manner: "In this problem, there is one dependent variable (T) and one independent variable (E). The relationship between the variables will be clarified by answering the following question, 'To what extent are the variations in T explained by the variations in E?' "

If more than one independent variable is hypothesized as affecting the variation of temperature (elevation or E and precipitation or P, for example), the hypothesized relationship is stated in the same manner: "To what extent is the variation of T related to the variation of P? Do the combined E and P variables provide a significant explanation of T variation?" Given the operational definitions of the variables (T, E, P), there is no doubt as to what the researcher intends to investigate.

Survey of the Literature

Although not always essential, it usually is necessary that the researcher include a preliminary survey of the literature in the work plan. He has made such a survey in order to weigh the feasibility of making the study. It will clarify his thinking to write a statement describing and evaluating the available literature pertaining to his study. A second value of such a survey is to give the advisor some information on which to base his guiding opinions and suggestions. In describing the literature, three points should be emphasized:

    (1)   the amount of work that has been done previously on the subject,

(2) an indication as to the strong and the weak areas of
     the existing literature, and
(3) the trends pertinent to the research problem as revealed
     in the survey of the literature.

Procedure, Tools, and Time Schedule

The major portion of the work plan is a three-step sequence
of planning: (1) the tasks considered necessary to complete the
problem, (2) the tools required, and (3) the estimated time required for their completion. The first step is to make a list of
individual tasks which must be completed. After this list is
drafted, the individual items can be rearranged, combined, and
divided into a reasonable sequence.

In the arrangement of the sequence of tasks, the order in
which they are performed is not necessarily the order in which
they are presented in the completed report. A map of the study
area may be on page one of the introductory chapter in the finished study, but it may be scheduled for completion with other
maps after the final draft has been written. As research proceeds,
rearranging may be in order; but it is necessary at this stage
that a tentative plan be envisioned.

After the succession of steps has been decided upon, the
steps and the particular tools needed for each one should be reviewed. Such tools may be *physical*, as a particular drafting tool,
for a specialized map, or they may be *conceptual*, as the knowledge
for designing a stratified random sample. At this point, the researcher will be able to make an inventory of his resources and
note which items are available and which ones must be obtained.
Again comparing the geographer to a builder, it is at this point
that an inventory of materials and needed skills is made so that
they may be ordered and be available at the time they are needed.

The third step in this sequence is to check each item and to
estimate the time required for each step. It is also advisable here
to make note of any expenses which will be incurred in addition
to the time investment of the researcher. In the event that either
time or cost exceeds the researcher's resources, some arrangement
must be made. This may involve modification, such as the elimination of a questionnaire for a personal survey. In some cases, it is
also possible to trade time for costs. If time is crucial, some part
of the work (perhaps typing of the completed paper or drafting

maps) may be accomplished by employing help. It is well to consult with the advisor if there is any question involving the legitimate work of the researcher and the work that may be done by others. Certainly, no researcher should claim credit for work done by others.

During this phase of the planning, some rearranging may become necessary. Perhaps a book that is needed in Step 1 must be ordered, and Step 2 can be shifted to Step 1 while the student is awaiting its arrival. A step may not be possible until a certain skill is acquired. If a course in sampling is needed or if field skills are required, the researcher may wish to take a course before completing his survey.

A question might arise at this point as to how the researcher can estimate the time required to finish a specific task. When it is not known, the best estimate must be made. For example, the book on order probably will take two weeks to arrive. The course in sampling will end on a certain date. The amount of time needed to make the map may be determined by comparing it with other maps made. The time to read the book can be estimated from the researcher's own knowledge of his reading speed and from the number of pages in the book.

It is not expected that time estimates will be perfect. Nevertheless, it is necessary to determine the general time requirements for various tasks and to formulate an overall timetable, or the researcher is courting disaster. Deadlines should be made in a reasonable manner, and it is good to allow for about 10 percent more time than anticipated. Without a time schedule, the "law" of Parkinson takes effect (the time needed to complete a task expands to occupy the time available for its completion). Having a deadline puts the researcher on the initiative and alerts the mind to prepare for the task ahead. It makes organic thinking possible, an essential requisite so important in research.

There are other points that should be kept in mind concerning a time schedule. Some tasks are easier and some more difficult than envisioned. If one step is completed ahead of schedule, the researcher should go on to the next. This is also related to the variation in the productivity of the researcher. He may have an extremely active day and wish to postpone a relaxation period in order to take advantage of his productivity. Related to this is the fact that all individuals have particularly productive times of the day. If it is really necessary to work thirty minutes cleaning the typewriter and locating paper, these tasks should be done during some period of little creativity.

Time Planning

Before assigning exact periods and dates of completion to individual tasks, the researcher should establish general categories of time to be invested in separate phases of the research. There are no hard rules established for this allocation of time, but some general guidelines will help the beginning student. As more experience is acquired, he may modify this schedule the better to suit his individual work tempo.

Most research work has two points rather firmly established —a starting date and a completion date. The starting date is usually immediate, and the completion date is determined by the college or university schedule, funding agencies, and other outside forces. Within these limits, the researcher sets his beginning and ending dates to fit his personal needs best. The length of this time period has some influence on scheduling; but as a rule of thumb, the work is performed in two stages. Perhaps in two-thirds of the time available, the researcher is involved with preparing the study, arranging, gathering, and analyzing data. Specific items include formulating the problem and hypotheses, selecting the title, surveying the literature, preparing the outline, making the tentative bibliography, collecting data, analyzing and testing data, and organizing notes under appropriate groupings.

The second phase, the remainder of the time, is assigned to the work of actually preparing the information found in the research for presentation in a written report. In order of time these steps are:

(1) a logical analysis of the subject matter preparatory to arranging it in the proper sequence for presentation,

(2) the outline for the completed report,

(3) the writing of the first rough draft of the report,

(4) the rewriting and polishing of the rough draft,

(5) the assembly of the final supportive materials (maps, tables, bibliography, etc.),

(6) the writing of the final draft,

(7) selection of a final title, dedication statements, index tables, abstracts, and other related activities necessary for final presentation of the study.

It is during the completion phase of the report that the final draft is typed. Some time must be allowed for it, whether someone is employed to do the actual typing or not. Also, there may be a need during the final stage to review Step 1 in order to fill gaps in the study, ones that became apparent during the logical analysis of material and the completion of the final outline.

After the major segments of the research have been assigned time limits, specific items may be assigned. The detail desired will vary greatly depending upon the type of study, the length of time involved, and the individual characteristics of the researcher. One would probably not wish to list the day he plans to read each book; but if a certain book is to be secured through interlibrary loan, the researcher must anticipate when the book it needed in order to place his order sufficiently well in advance so as to complete its use by the time its return is required. In this regard, it is definitely advisable to plan the points at which significant works should be obtained and completed. It is also advisable that the researcher keep a diary of this daily progress. Much research has been accomplished with little detail in planning; but the researcher who organizes his work within the structure of the total project will be able to meet critical deadlines, and his final product will reflect his careful planning.

Each research study is unique, and time requirements for each task will vary. However, assuming that the research project is scheduled to be completed within the temporal framework of a college semester (approximately seventeen weeks in duration), the student researcher might construct a work/time master schedule based on the best estimates he can make (see Table 1, Page 26).

It is evident that Table 1 is highly generalized and that it should not be considered as inflexible or rigid. Rather, it should be looked upon as a sample or guide that indicates the general magnitude of the various tasks comprising a research study, a guide which the student should consider in constructing a master schedule to fit his particular requirements.

In the event that the research study represents a major piece of work, such as a Master's thesis, it is most likely that a longer period of time should be allocated to writing the report (Phase II) than is shown in Table 1. Often, research projects take longer than one semester, but the percentage of time devoted to various tasks most likely will remain about the same.

TABLE 1
Sample Work/Time Master Schedule

Task	Proposed Time
Phase I. Planning Actual Research	12 weeks (total)
A. Survey of Pertinent Literature	1 week
(identification of significant professional papers, reports)	
B. Acquisition of Pertinent Source Materials	1 week
(base maps, statistical reports, surveys, etc.)	
C. Collection of Data	6 weeks
(library and field research)	
D. Analysis and Processing of Data	4 weeks
(compilation, evaluation, selection of data)	
Phase II. Preparing the Written Report	5 weeks (total)
A. Preparation of Report Outline	½ week
(outline of chapter and major headings)	
B. Preparation of First Draft	1½ weeks
(compilation and first writing of complete report)	
C. Revision of First Draft	1 week
(rearranging, editing, modifying, rewriting)	
D. Assembly and Preparation of Final Supportive Materials	1 week
(drafting maps, illustrations, tables; assembly of bibliography)	
E. Preparation of Final Draft	1 week

acquisition
of relevant data

Acquisition of geographic data involves several techniques that have been well established for many years. All major departments of geography offer one or more courses focusing on these techniques and the development of associated skills. In fact, these skills are considered essential to the professional training of a geographer. They are often referred to as "techniques" or "methods" courses, for example, Field Techniques, Cartographic Techniques, and Quantitative Methods.

Whether these skills are referred to as methods, techniques, or tools is not particularly important. They are relative terms, and in a hierarchy of importance, the order would be: (1) method, (2) technique, and (3) tool. In this discussion, the term "method" will be used to describe the general research scheme or framework which determines the kind of result sought—a law, a norm, or a history. The type of research problem determines the "techniques" or skills which must be applied to produce the solution. A technique refers to a design of procedure which best performs the job at hand. For example, a geographer might employ "cartographic techniques" to illustrate the spatial distribution of phenomena. A "tool" is an instrument or device used for skillful technique operation. The geographer uses a wide variety of tools, depending upon the specific nature of the task at hand. Tools include aerial photos, weather instruments, maps, computers, drafting instruments, alidades, and compasses. In addition to such physical tools, mental or conceptual tools, such as a random sample or a mathematical model, are also used.

Like other disciplines, certain basic skills are used in geography, such as library research, logical deduction and induction, and written expression. However, these skills are so much a part of, and so essential to higher learning that they are taught by departments established for no other purpose.

Special skills, unique to geographic research, are usually taught as geography courses and require considerable time and direction for their mastery. This is true of field techniques, cartographic techniques, air photo interpretation, and quantification. Research work in geography implies that a working knowledge of all these skills has previously been attained, as well as a mastery of one or more techniques pertinent to the specific research task at hand. Frequently, the beginning research student faces some confusion in applying these skills. The courses are generally taught as isolated units. They are taught as skills, which they are, and it is often not clear to the student as to when and where a specific technique should be used within the overall structure of the research process. The purpose of this chapter is to establish the order of the research process and the techniques most effective to each stage.

Scientific research is the gathering and processing of data on which scientific truths are based. It is a logical and systematic sequence of related steps. The order of these steps, or stages, is: (1) the collection of data, (2) evaluation of the data, (3) analysis of the data, and (4) prediction based on the analysis. Each stage of the research must be pursued in the proper order, since it is virtually impossible to complete any stage without the knowledge obtained in prior stages.

The ultimate aim of scientific research is to build theory as a basis of understanding and prediction. It might be noted that some research, especially that of an exploratory nature, does not predict but, rather, obtains basic new information which was previously unknown. In situations where data is lacking or difficult to obtain, the gathering and organization of the material may be a major task. Usually, elementary research at the undergraduate level entails only a descriptive report. Most graduate research work, however, and especially a thesis, requires analysis and some form of model development or prediction. Geographic techniques provide the necessary skills for the competent completion of each of these four stages.

Geographic techniques are employed in all stages of the research process. Each specific technique is employed most intensively in one or two stages; and it is well to recognize the contribution of a given technique when planning the overall research design. Table 2 indicates the technique for each stage of the research, although they are not necessarily limited to the particular stage shown.

It is unlikely that all techniques shown in Table 2 will apply to a specific research problem. But science attempts to find truth with "no holds barred," and the researcher may employ whatever skill he finds useful in his search for a true answer to his problem.

Data collecting is essentially making an inventory of pertinent facts. Some of this factual information may be in the form of printed articles, books, maps, and published tables, charts, and

TABLE 2
Geographic Techniques Used at Various Stages

Stages	Techniques
I. Data Collecting (Gathering pertinent information.)	A. Library: search and survey for existing relevant published materials.
	B. Cartographic: analysis of existing maps, air photos, and imageries.
	C. Field: acquisition of new data; mapping and interviewing procedures.
II. Evaluation of Data (What is the precise nature of the information gathered?)	A. Appraisal of Published Materials: type of information, quantity, source date, quality, and relevancy.
	B. Evaluating Cartographic Materials: total areal coverage; type of phenomena illustrated: scale, date, accuracy, quality, and relevancy.
	C. Evaluating Field Data: type of data; how obtained; intensity, scale, accuracy, and quality.
III. Data Analysis (What does the information mean?)	A. Logical Procedures: deduction, induction.
	B. Statistical Compilation: arrangement of data into tables, diagrams, and models.
	C. Cartographic Compilation: arrangement of data spatially; construction of maps and space-distance models.
	D. Correlations and Relationships: determination of causal and non causal associations, correlations.
IV. Development of Theory or Prediction (What are the results? Can they be projected?)	A. Construction of Models, Laws, and Norms: development and validity of basic models; mathematical, cartographic, logical documentation.
	B. Projections: predictive accuracy to specific areas, regions, world; temporal restrictions.

statistical data. If so, it is obtained by reading, taking notes, comparing maps, and studying image reproductions. If it is not in published form, the information may be in the "field," thus requiring the utilization of one or more field techniques. This may involve mapping selected phenomena of the visible landscape, either singly or in combination; taking photographs; making notes; and either directly or indirectly interviewing people who have the desired information. The researcher, after collecting the essential data, must appraise and organize it in some logical order so that it may be analyzed in relation to the hypotheses at a later date.

Library Resources

Since the very term *research* denotes a process of checking and rechecking in order to become certain, it is evident that the library occupies an important position in the process. Probably the advanced geography researcher relies less completely on library sources than does the beginner. Presumably, the former has a better background in the literature of his research topic; he emphasizes individual reasoning to a greater extent in his work; and he generates more of his own data. Even so, both the advanced and the beginning student find the library essential for reviewing the literature and obtaining source material for their studies.

There is no standardized first step for beginning the actual search for material which has been written on a geographic topic. However, the library is the depository for recorded information and is a logical starting point. The elementary and background material already known about the topic may be concisely stated in the encyclopedia, an atlas, or a statistical almanac; and the research should probably begin by reading these sources. In addition to acquiring background information, a start is made in developing the reference bibliography.

The next step in acquiring information in the library is to survey the card catalog. Here all books are indexed by author, title, and subject. After noting any potentially helpful items in the subject file, the researcher should check authors and titles for any items which he may wish to use in his bibliographic file.

The third step is to search the standard and special references in the library that are available to researchers. In the first category, *Reader's Guide to Periodical Literature*, *Guide to Reference Books*, *Basic Reference Sources*, *Cumulative Book Index*, and the

Library of Congress' *National Union Catalog* are suggested. For geography researchers, the indexes to each of the major geographical periodicals should be searched carefully. Documents focusing on comprehensive bibliographic references are available; these include existing lists of reference materials (see Appendix A). For articles in foreign languages, one must refer to the specialized references of the selected area.

The student researcher should make bibliography reference cards of all pertinent published material. In making these note cards, he should list complete information. It is from these cards that the final material for the printed bibliography is obtained. The cards should appear much like the examples below. The space following the bibliographical data is for annotation comments concerning the material in the publication. It is recommended that the cost of the publication be included when available.

Book Example:

```
        Geographer, George. 1970. Geography of place.
        Dubuque:  Wm. C. Brown Company
        Publishers.
                    (Annotation)
```

Periodical Example:

```
Space, Thomas. October, 1970. Political voting patterns:
a model for political geography. The Spatial Enquirer,
pp. 47-55.
                (Annotation)
```

The bibliography will expand as reading for substantive material proceeds during the research period. All scholarly articles normally refer to other works which shed additional light on the subject. In fact, such references are so numerous that the researcher will be thankful that he has a clear and well-defined problem. Otherwise, the amount of material referred to is so large that it would be impossible to determine which leads are worth pursuing and which ones should be ignored.

NOTE TAKING

   In addition to the bibliography reference card, the researcher must record information on note cards from which he eventually writes the report. Proper note-taking is absolutely essential for the successful assembling of data for the research project. It begins in earnest immediately after the working outline is completed, and it continues until final organization for writing the research report. As the researcher searches for the specific ideas, facts, and statements that bear on the problem, he makes a note of each significant item found. In general, he seeks to record all useful items and avoids taking notes that do not contribute directly to his specific study.

   There is no one system of note-taking, but every researcher should have some system which is meaningful to him. The card itself may vary in size from three-by-five inches to five-by-eight inches, depending on the desire and the writing style of the user. If the researcher does not have a format, he should refer to a basic text in research for a review of methods used. In any event, four items must be known concerning any substance note—the author, title, date, and page of the reference. This does not imply that all of these facts must be shown on the card. For instance, "Doe, *Research Methods*, 17," will lead the researcher to the proper bibliography card if only one work by Doe was used as a reference. Also, "Doe, 1970, 17," is sufficient. Some researchers number their bibliography reference cards as they use them and then record that number on the substance note card. If the book on research methods by Dr. Doe was the third one read, a number "3" would be placed on the bibliography reference card. Then the item recorded on the note card could be labeled "3, 17." In writing the report, a reference to "3, 17" would direct the writer to the work of Dr. Doe, and it could then be properly footnoted. It should be kept in mind, however, that the bibliography reference cards must have the complete name of the author, the title, publisher, publication location, and date of publication.

   The material on the note card should be carefully selected. The researcher should never put more than one item on a card. The card should be made either at the particular time an item worthy of note is read or at the completion of the reading of the entire work. This last method saves some worthless note-taking and interruption in reading. If it is used, some method of indicating the location within the article of the important item to be recorded must be devised.

Notes may be categorized into five groups: (1) *the précis,* a useful recording of a thought in the researcher's own words; (2) *summary notes,* condensations of pertinent information in the article; (3) *paraphrase notes, similar to a* **precis** but more detailed; (4) *critical notes,* which evaluate and make critical comments on the material; and (5) *quotation notes,* to be used when a statement is so appropriate that it is wanted for inclusion, and the researcher cannot devise a better way to state the thoughts in his own words. Needless to say, great care is required in taking quotations accurately and reproducing them in the paper in proper context.

Reference footnote form varies, and the researcher should follow the format designated by his department or institution. A common form is to place a number at the completion of the material to be noted (raised one-half space) and to show the reference at the bottom of the page in the same way that footnotes in this book have been treated. A shorter method, one which is more efficient and which is becoming widely accepted, is the inclusion of the reference immediately following the material to be noted. The following is an example: (Geographer, 1970, 21). The important consideration is that the reader of the research report be given sufficient information regarding the reference in order to check it readily in its original form. The information tells him who said it and when. He may locate the source from the bibliography and, if he wishes, investigate the quality and reliability of the reference.

APPRAISAL OF LIBRARY SOURCE MATERIAL

Judging or evaluating source materials is part of the training of the research scholar. The value of the completed research work depends in large measure on the quality of the sources used.

Source materials may be classified as primary, secondary, and tertiary. Often, a particular work does not automatically fall into one category or the other but varies in relation to the nature of the study. For example, a textbook is usually considered a tertiary source since it is compiled mostly from secondary sources; but it may be a primary source in a study in which the major objective is to determine how textbooks treat the subject of model building.

Primary sources are usually considered the best sources in research. Such source material is near its original form and is relatively free of editing, alteration, or modification. As such, it tends to be divorced from external influence, judgment, and bias of others which might lead to unsound interpretation by the researcher. Primary sources, then, are original descriptions or analyses of a process or event. In this category, one commonly finds data concerning experiments, interviews, questionnaires, field studies, letters, diaries, autobiographies, creative works, and statistical reports. No interpretation by the collector of these sources is attempted.

Secondary sources are usually factual accounts written about a subject. The information commonly represents selected data taken from primary sources that have been organized and interpreted by the writer. A primary source may contain some secondary source material. For instance, a field study may include material interpreted from previous field studies and analyzed by the field reporter. This analysis then becomes a secondary source, as it includes a judgment factor. Only new information directly attributed to the field researcher's own efforts is primary in nature.

Tertiary works are compiled from secondary source material. Most textbooks are in this category. Well-conceived tertiary works become widely accepted and serve as standard reference documents.

A source may be primary, the interpretation absolutely correct—and still be of little value. Its use in such cases would be an indication of poor scholarship on the part of the researcher who is judged on his critical evaluation of each source cited. Obviously, any position could be supported by citing someone. For example, a recent political analysis in a national news magazine directly quoted twelve sources. This list of implied political "authorities" included an aluminum plant worker, a printer, a boxing manager, two pollsters, an unnamed "moderate Republican senator," and a "White House advisor." Two glaring research errors are obvious: lack of precise identification of the "authorities" and lack of any indication of their competence.

The research student must also exercise discrimination in the quality of the sources; this is based on his ability to judge each work. Some of the criteria to be applied in such judgment is external. In this group, such things as the reputation of the author and the publisher, and the comments of critical reviews and annotated reference works are considered.

Internal criteria require more discriminating skill. The researcher will wish to arrive at some evaluation of the quality of the work before investing too much time in it. The material's usefulness may be evaluated by reading the introduction, the purpose, or the preface to determine if the intent of the author is to treat the subject in a way that appears satisfactory. The date is evidence of the chronological pertinence of the study, and any comments about the author indicate something of his reliability. A sample reading of a few pages will suggest the level of reasoning in the work, as well as the clarity of the writing. At the same time, footnotes can be checked for accuracy and adequacy. The bibliography of sources used, whether lengthy or short, is an indication of the quality of the work.

PLAGIARISM

The "coin" of the scholar is his scholarly creations. Plagiarism is the use of these creations without giving credit to and/or getting proper permission from their creator. Such creations include not only the exact words of the originator, but also his ideas, phraseology, and original organization of materials. Such originals are protected by law; and in serious cases of usurpation, legal liability can be established.

Most plagiarism is unintentional and results from errors in note-taking or reporting. Even so, such errors are not to be treated lightly because they indicate lax scholarship. It is not difficult to avoid this pitfall if extreme care is taken to give proper credit. When any author's copyrighted material is used, there is a possibility that this use will compete with the original, and so permission should be obtained. The best single criterion to keep in mind is: "Does use of the material impair the value of the original?" Such interpretation may be difficult, and if too strictly followed, could preclude the use of almost all sources. The writer should use reason, and when in doubt, risk erring on the conservative side. The length of the material used is not a criterion for determining plagiarism. Each situation differs, and there is no rule of thumb to serve as a guide in this matter. Most types of creative work may be protected by a copyright. A potential author should be familiar with the regulations.[1]

1. United States Copyright Office, *Copyright Law of the United States of America* (Washington, D. C.: U. S. Government Printing Office, 1967).

Field Techniques

In many research problems, key information or essential data may not exist in published form. This is particularly true if the research area is small and detailed data are required. It then becomes necessary to obtain information by employing one or more field techniques. Geographic field techniques essentially consist of recording direct observations within a well-conceived and scientific framework. They provide a structure that enables the researcher to make order out of chaos.

As is true in other sciences, the techniques and research equipment used by geographers to collect raw data have evolved over the years. Direct observation as a major source of data is deeply ingrained in the history of geographic thought. Indeed, it was less than 100 years ago that a "geographer" was one who journeyed to exotic regions and returned to recount his travel experiences, including general observations he had made. In the event that his descriptive report was presented before a prestigious group, such as the Royal Geographical Society of London, the geographer had reached the apex of his career. Today, direct observations can be accurately recorded and scientifically structured. Many geographical studies lack sufficient and basic data without the information provided by modern field work. Field work is the collection of raw data by primarily using: (1) mapping techniques, and (2) interviewing techniques. Most field work employs both these techniques because one implements or supplements the other.

MAPPING TECHNIQUES

Mapping raw data is concerned with the spatial recording of visible features, or phenomena, of the landscape, features that are capable of being categorized. Field mapping consists of several distinct steps which must be completed in order. These steps are: (1) determination of the *scale* of the mapping task, (2) *classification* of the data to be mapped, (3) selection of the *base map* upon which the data will be recorded, and (4) the *actual field mapping*.

The selection of the scale determines the size of the minimum mapping unit. If the research matrix is small and detailed data is required, the minimum mapping unit may be only a few square feet in size. On the other hand, if the research problem is concerned with a large matrix and data is less detailed in nature, the minimum mapping unit may be several acres in size. The scale of mapping determines, to a large degree, the way in which data is classified and the precision of the actual field mapping.

A map is a man-made representation, drawn to scale, of a segment of the earth's surface. It is always smaller than the part of the surface which it represents, and so it is not possible to show each and every aspect of the total landscape exactly as it occurs. Certain features, or phenomena, must be selected to be recorded, and those selected must be generalized, or classified, in some meaningful manner. The process of classifying, or categorizing, data requires serious thought. The type of classification bears directly on the outcome of the research problem. The need to classify and categorize is not unique to the field of geography. In all sciences, physical or social, the development of taxonomies, or classifications, is prerequisite to standardized research. The orderly accumulation of information leads to the formulation of conceptual frameworks and basic laws. The classifications of elements, plants, animals, and rocks, as well as of human activities and events, are a few examples of a multitude. The existence of many recognized subfields within the discipline of geography requires the general geographer to have a working knowledge of many taxonomies, and the research geographer must have the ability to conceive sound and scientific classification systems as the need arises.

Mapping implies recording facts, drawing boundaries, and making notes on *something*. This something must have reference points or control lines accurately placed on it so that the mapper knows where he is at a given time and can accurately locate features of the landscape on his field or base map. A base map may be looked upon as being a type of notebook upon which facts and observations are recorded. However, unlike a notebook, the observations, or facts, not only can be recorded, but their spatial or areal extent can be accurately defined. If the classification system of the data being collected is sound, the completed map becomes a scientific document capable of showing a wide variety of pertinent information with many uses and applications. A finished map is essentially a table that presents data spatially.

There are many types of maps and map-like devices that are widely used as base maps for field mapping. They include vertical aerial photos, topographic maps, plat or cadastral maps, air charts, and even a sheet of paper on which some control or reference points have been placed. The selection of the most appropriate base map depends upon the type of data being collected and upon the scale or size of the area being mapped.

Aerial photos are widely used today as base "maps" because they are camera images of a part of the earth's surface. All visible features are shown—crops, houses, trees, gullies, roads, and so on. The aerial photo in itself provides a great deal of information, such as the areal extent of cultivated land, landforms, forests, settlement, and transportation patterns. An experienced researcher may compile land use data and construct a land use map from the photo itself, using interpretation techniques verified by spot checks in the field. A photo is not really a map—it is a picture of all aspects of the landscape—no selection process has taken place. Conventional photographs show everything, although filtering devices and recent techniques of remote sensing allow some selection of features to be reproduced.

Photographing and interpretation of aerial photos is essentially an art that has been developed since World War I. As a geographic research skill, it attracted little attention until 1943. At the present time, in addition to conventional photos, geographers utilize infrared, radar, and other forms of remote imagery. These devices are of major concern to the research geographer because they provide a mechanism to obtain and describe data.

Oblique aerial photographs are taken when the optical axis of the camera is inclined from the vertical. Oblique photos, when used singly, distort surface features and are of little value for mapping purposes where true area and shape of features are needed. Vertical photos, on the other hand, are those taken by a camera pointing vertically downward, and they illustrate areas and ground objects as they actually exist. If, however, the camera is tilted slightly from the vertical, or if there is a great deal of topographic relief, some distortion will be present. In all cases, the edges of the photo are distorted to some degree, and only the central portions of a vertical photo truly represent the surface of the ground. However, despite these disadvantages, vertical photos serve as excellent base "maps" because of the detail shown and the large number of control points which are not normally present on other types of maps. Most parts of the world have been photographed one or more times by government agencies or by commercial concerns. Every county in the United States has photographic coverage, and photos are available in various scales (see Figure 1 and Appendix A).

Topographic maps are also widely used as base maps. They are large-scale maps and are available for many parts of the country. They show the exact locations of specific physical and cultural features which serve as reference points in field mapping. Like all maps, they do not show all of the landscape, as certain features have been selected to be illustrated, while other features have been omitted (see Figure 2).

In collecting urban land use information, plat maps are often used. These maps are concerned only with property boundaries and show the dimensions of properties and lots in feet. They are available at large scales in most urbanized areas of the country. Inasmuch as distances are precisely given on the maps, compilation of the square footage of each type of land use is easy to compute (see Figure 3).

In the unlikely event that no maps of the research area exist at a scale necessary for field mapping purposes, the researcher, by using simple and inexpensive equipment, may construct his own base map. A piece of paper, a compass, some type of sighting device (alidade), a table with tripod, and a tape measure are all that are required. The researcher then can construct a series of points and lines, each in its proper direction and distance from the others, which will serve as a controlling framework within which specific phenomena may be accurately located.[2] Plane table mapping is rarely used, as base maps for most areas are available. However, in microarea studies, plane table mapping may be necessary (see Figure 4).

After an appropriate base map has been acquired and the type and classification of data to be obtained have been determined, the researcher is ready to begin the actual field mapping. Obviously, the mapper must be able to locate ground objects precisely on his base map; and if a sufficient number of features may be identified on his base map, he may then draw the boundaries around whatever phenomena concern him. It is strongly advised that, when gathering raw data, the student researcher collect more than he thinks is pertinent at the time, and that he utilize the most detailed system of data classification that is possible at a given scale. It is not an uncommon practice during the actual field mapping process to identify additional pertinent data that were not recognized previously. Often, the value of this additional information is not fully realized until the data are analyzed at a later stage of the research problem. Multifeature mapping is also strongly recommended, as the resulting information may

2. For detailed information concerning plane table mapping see: David Greenhood, *Mapping* (Chicago: The University of Chicago Press, 1964), pp. 203-239.

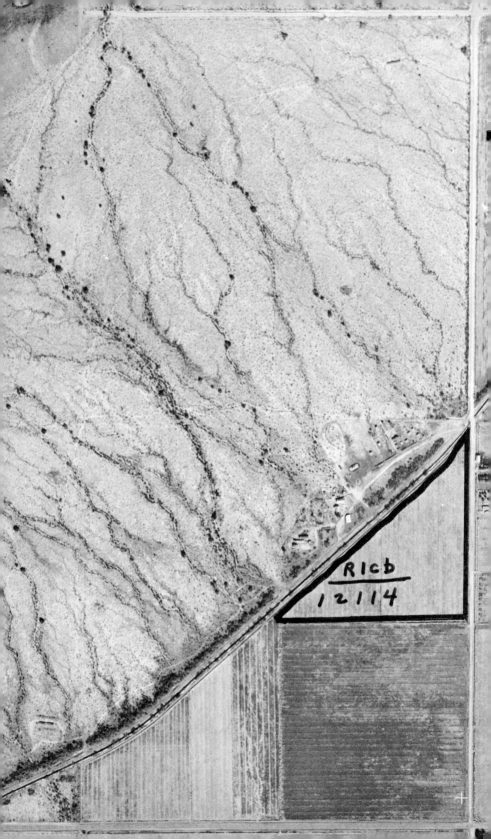

R1cb
12114

Figure 1. A portion of a vertical aerial photograph (scale: 1 inch = 1,000 feet). Note the presence of field patterns, small topographic, vegetative, and cultural features that are not normally shown on maps. Aerial photographs for all parts of the country may be obtained from the United States Department of Agriculture, Agricultural Stabilization and Conservation Service (ASCS), Washington, D.C. 20250. Each county ASCS Office has an index of the photographs giving the code numbers and date of flight for individual photos in the country. These code numbers must be obtained, as well as an order blank, before photographs can be purchased.

Land use mapping may be easily accomplished using vertical aerial photos as base maps. The example below illustrates multifeatured mapping utilizing the fractional code system (details concerning the classification system of land use and physical features used in this example may be found in Appendixes B and C).

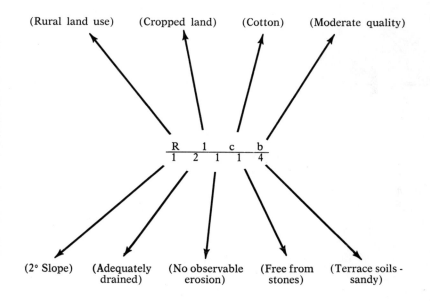

(Rural land use) (Cropped land) (Cotton) (Moderate quality)

$$\frac{R}{1} \quad \frac{1}{2} \quad \frac{c}{1} \quad \frac{b}{1} \quad \frac{}{4}$$

(2° Slope) (Adequately drained) (No observable erosion) (Free from stones) (Terrace soils - sandy)

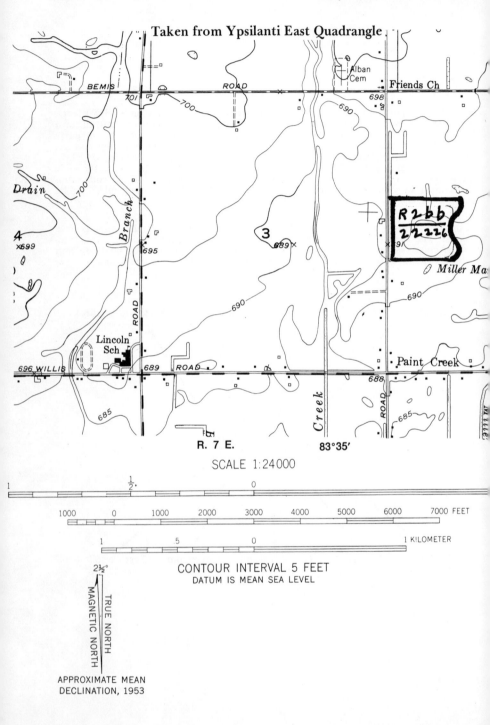

Taken from Ypsilanti East Quadrangle

SCALE 1:24000

R. 7 E.   83°35′

CONTOUR INTERVAL 5 FEET
DATUM IS MEAN SEA LEVEL

2½°

TRUE NORTH
MAGNETIC NORTH

APPROXIMATE MEAN
DECLINATION, 1953

Figure 2. A portion of a topographical quadrangle, 7½'x7½' Series, Scale: 1 inch = 2000 feet (Courtesy of the U. S. Department of Interior, Geological Survey). Note the absence of small ground features found on a vertical aerial photo. However, topographic maps show political boundaries, systems of coordinates, and they identify major cultural and topographic features which may be used as control points in field mapping. The area outlined on the map illustrates how topographic maps may be used as base maps for field work. The fractional code system of mapping and the classification systems are explained in Appendixes B and C.

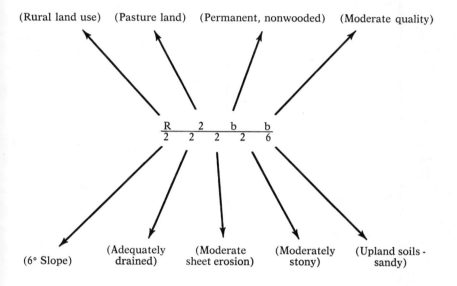

These types of topographic maps are made by the United States Department of Interior, Geological Survey, Washington, D. C. 20242. To order maps, and for information on coverage, scales, dates, and prices, write to:

> For states east of the Mississippi River:
> Distribution Section, Geological Survey
> 1220 South East Street
> Arlington, Virginia 22202

> For states west of the Mississippi River:
> Distribution Section, Geological Survey
> Federal Center
> Denver, Colorado 80225

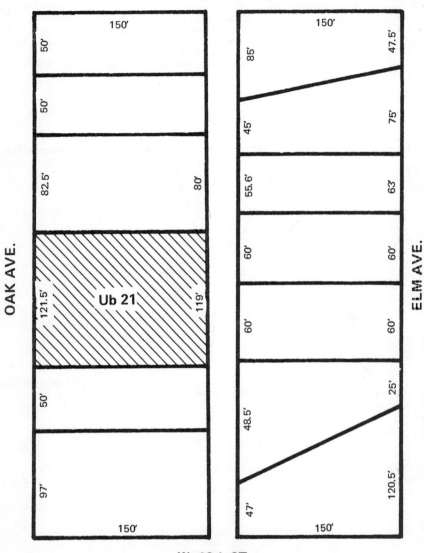

Figure 3. A portion of a plat map. These maps are often used as base maps for urban land use mapping. Property boundaries and exact distances are shown, and the precise square footage of each type of land use can be easily computed. Plat maps may be obtained from City or County Government Offices. This example illustrates a two-family residence in moderate condition, constructed since 1960 (see Urban Land Use Classification, Appendix C).

throw light on unexpected areal associations. When the researcher is collecting raw data, it is not always possible for him to antici- pate in advance the value of the data and the new concepts or hy- potheses which might evolve.

The student who engages in field mapping is well advised to follow a carefully detailed plan of action. Field work is a time- consuming research technique. However, a well-conceived field plan will prevent much wasted effort and time. Fundamental to good planning is: (1) a clear statement of objective(s), (2) a care- ful pre-field study of existing documentary materials, (3) the prop-

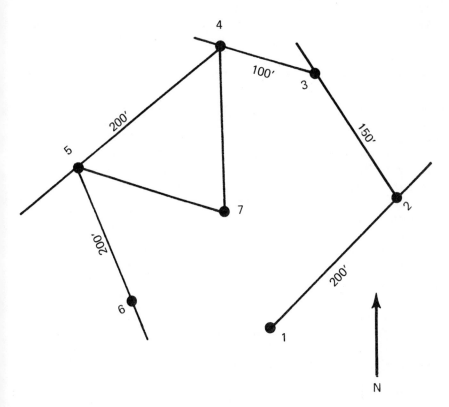

Figure 4. A sample plane table base map (scale: 1 inch = 100 feet. A series of control points has been established, as well as a line of traverse. From this frame- work, other control points may be accurately located and the areal extent of phenomena mapped. From each station of observation, the exact direction and distance to a subsequent station must be determined precisely if a base map is to be accurate. From *Basic Statistics* by William Hays, © 1967, Wadsworth Publishing Company, Inc., Belmont, California 94002. Reprinted by permission of the Publisher, Brooks/Cole Publishing Company.

er preparation of base maps and related materials, (4) a reconnaissance of the research area, (5) the field mapping, taking extreme care to record data accurately, and (6) a post-field study to ink in penciled lines, check for gaps, assemble field maps, organize field notes and on-site photographs, and evaluate any unexpected information that has been revealed.

INTERVIEWING TECHNIQUES

In many research problems, raw data, which may not be visible at the time, are needed and, therefore, they cannot be mapped in the field. For example, in an agricultural study information concerning the marketing of products, the researcher must know facts about rotation of crops, types and amount of fertilizers used, labor requirements, and other related variables if a complete understanding of the agricultural geography of the area is to be acquired. For data of this nature, the researcher must resort to interviewing techniques. Assuming that an excellent base map is used and that the classification system of data is clear and concise, data derived from mapping techniques have a high degree of accuracy. It is often not possible to achieve such accuracy by employing interviewing techniques. The individual interviewed may not provide the researcher with highly accurate information because: (1) he may not fully understand the question; (2) the researcher may not fully comprehend the answer; (3) the person interviewed may either intentionally or unintentionally provide misleading information; or (4) the questionnaire may not be properly structured so as to elicit unbiased responses. Nevertheless, these inherent difficulties may be minimized if proper safeguards are taken.

The interview may be conducted directly, or a questionnaire may be distributed through the mails or by some other method. The type of information gathered, and its uses, are generally the same whether obtained directly or indirectly. Both require considerable skill in structuring questions in such ways as to obtain data that may be interpreted correctly.

The objective of a questionnaire is to gather factual data from the respondent when no other source for obtaining the required information is available. The questions should be formulated in such a way that the responses reveal quantitative or qualitative information. Ideally, the responses should be in such a form that tabulation can be done without judging or assuming what answer was intended by the respondent (see Appendix D).

Direct conversation as a means of data gathering is most useful for obtaining qualitative data, although specific data may also be obtained directly. The interviewer should have the questions clearly in mind and properly worded so that he need not refer to any notes or write down any responses. This is considered desirable because careful note-taking, usually valued so highly in research, interferes with candid answers on the part of the interviewee. For best responses, a relaxed, informal, spontaneous atmosphere is desirable. It is sometimes necessary to gather specific data by the interview method where exact questions and note-taking are essential, although seldom would this be necessary in introductory research problems. For anyone contemplating interviewing, it is recommended that he consult a more complete work on interviewing.[3]

The mailed questionnaire has fallen into some disrepute in recent years because of the quest for mass information and the availability of the mails. As a result, those who receive such requests tend to disregard them. It is, therefore, all-important to cultivate the receptivity of the questionnaire respondent in any way feasible. One good list of suggestions includes ten things the researcher can do to solicit a reply:

(1) ask only for information not available from other sources;

(2) ask only important and significant questions;

(3) try to engender a desire within the respondent to answer questions honestly;

(4) do not ask questions which are to be answered objectively but which imply further explanation;

(5) do not promise the respondent a summary of the questionnaire unless the researcher fully intends to keep that promise;

(6) try to develop a reason for the respondent to answer;

(7) use a questionnaire which is not too lengthy;

(8) encourage the respondent to sign the instrument;

(9) do not use humor or personal reference; and

(10) send an explanatory cover letter with the questionnaire but unattached.[4]

3. Margaret Brown and Elizabeth Sidney, *Skills of Interviewing* (London: Tavistock Publications, 1961).
4. J. Francis Rummel and Wesley C. Ballaine, *Research Methodology* (New York: Harper & Row Publishers, 1963).

There are several good references on proper questionnaire construction in the area of educational and social research. The student planning such a research tool should read one or more of them carefully (see Appendix A).

SAMPLING PROCEDURES

When using interviewing techniques to collect raw data, the researcher may find that because of the large size of the research matrix, the total population cannot be interviewed in a reasonable period of time. Obviously, total coverage of the research area is most desirable; but if this goal cannot be achieved, the researcher must devise a system of sampling.

Sampling procedures are less commonly employed in field mapping. The specific nature of the research problem and the type of data to be collected will largely determine the type of sampling structure. For example, if the researcher were attempting to determine the average price paid for a pound of tobacco to farmers in an homogeneous tobacco-farming area, he might structure a sampling procedure whereby one farmer in each group of ten would be interviewed. In his field work, the researcher would not interview each tenth farmer he encountered; but by using a table of random numbers, it would be determined, perhaps, that he interview the second farmer he encountered in the first group of ten, the seventh farmer in the second group of ten, and so forth (see Appendix F).

If, however, the data desired were more complex, such as determining the attitudes of the people concerning a proposed master plan for a city, he would need to ascertain a true representation or cross section of the population. He might then classify persons as to income group, level of education, place of residence, and length of residence in the city. For each group classified, he might then determine to interview one in twenty, again using a table of random numbers. Opinion polls and marketing surveys are examples of types of stratified samples. The accuracy of the data obtained from such a sample is directly related to the way the groups were classified and to the percentage of persons surveyed within each group.

In developing a sampling system, the researcher must take the following into consideration: (1) the type of data he wishes to acquire, (2) the degree of accuracy he wishes to attain, (3) the importance to his study of an area cross section, (4) the importance to his study of an economic, social, or cultural cross section of the population, and (5) the application of the data obtained. There is no general rule for devising a standard sample, as it would vary from one specific research problem to another. The student researcher who plans to develop sampling procedures is advised to study recent references on the subject.[5]

5. Leslie J. King, *Statistical Analysis in Geography* (Englewood Cliffs, N. J.: Prentice-Hall, 1969).

chapter **5**

analysis of data

Among the uses of the scientific method are analysis and prediction. It is at the analysis stage of research that the geographer weighs, compares, and tests his spatial hypotheses for acceptance or rejection. If it is found that the actual location and distribution of the variable under observation corresponds to that stated in the hypothesis, he accepts the hypothesis. The measurement of the extent of the similarity of pattern between the hypothesis and the "real world" must be determined by logical deduction or induction with the aid of statistical and cartographic techniques of analysis. On the basis of such tests, laws or norms may be established—at least a norm for the limited universe enclosed by the bounds of the research project involved.

Logical analysis by means of verbal symbols is a common process in science as well as in everyday reflective thinking. In this process, the procedure is guided by the laws of logic, and the results are expressed in words designed to convey accurate meaning to the recipient. Geographic research has always depended in large measure on the researcher's ability in verbal analysis. It is impossible to imagine a time when this will not be an important part of research work.

Evaluation of Data

The first step in analyzing the collected data consists of organizing, compiling, and comparing one set of data to another, and evaluating. The researcher, in evaluating the acquired data, must make judgments as to: (1) their relevancy to his particular study, (2) the total quantity, and (3) their quality.

With the statement of his problem firmly in mind, he must make judgments for each set of data. This might include a rank-order classification. The amount of data collected must be totaled, and the researcher may find that he has accumulated an excessive amount of data concerning a specific aspect of his problem, 'or, on the other hand, that he lacks sufficient information concerning some part of his study. In the latter event, he obtains the needed specific data by returning to library sources or to the field. Usually, this requires obtaining a small amount of very specific information dealing with one or two points.

Finally, the data must be evaluated in terms of their quality or accuracy. In the case of published materials, the researcher must consider the source—who wrote the article, how was the information obtained, and on what basis were the conclusions or predictions made. He should also note the dates of the articles or books consulted, the kind and amount of illustrative materials, and the bibliographies. In regard to existing aerial photos and maps, consideration must be given to the extent of the research area covered, the type of phenomena illustrated, the scale, date, and accuracy, that is, how a given map was produced and what sources of information were used in its construction.

The researcher must judge the accuracy of his own field work impartially, and he must determine its validity for documentation purposes. The evaluation of data most likely will result in the selection of certain data that prove to be essential and basic to the research problem, to the rejection of some information because of some inherent weakness or unsuitability, and to the setting aside temporarily of other data to be reevaluated at a later date.

Cartographic Analysis and Presentation of Data

After the significant data have been identified and isolated, the researcher can use any of a wide variety of cartographic devices to present the data in a manner which makes analysis most effective. Most of these cartographic procedures involve precise measurements, and they overlap with statistical or quantitative devices. It is not always possible, nor is it necessary, to make sharp distinctions, since cartographic analysis supplements statistical analysis and vice versa.

Since geography is the science of spatial analysis, it is likely that maps of some type will be included in any research study. A well-constructed map is a scientific document like a graph, a table, a diagram, or a model. Like these devices, the most appropriate types of maps will vary from one research problem to another.

In general, there are two major ways to present data on maps to preserve the true shape and area of phenomena. One method is the use of dots with numerical values assigned to each dot. Dot maps are commonly used to show population, production of commodities, and other phenomena. The other device is the use of isopleths, or isolines, or isograms. The prefix "iso" means *equal* and describes lines connecting points of equal value. For example, an "isotherm" runs through places of equal temperatures; an "isochrone" expresses equal amounts of time such as might be involved in travel or growth processes. "Isopleths" may be used, not only to show actual existing phenomena, but also tendencies. An "isallobar" connects places which have the same tendency of air pressure change within a specified period of time.

The researcher must decide what type of cartographic devices is most significant to his particular study. Maps can be constructed to show the spatial distribution of qualitative or quantitative (or both) aspects of phenomena. They may be single-featured, showing the distribution of one datum, multiple-featured showing the spatial distribution of several data, or they may show the ratio or relationship between two or more data. For example, a map might be constructed to show by dots the location of all aluminum-producing plants in the United States. This type of map would show the distribution of only one item of qualitative information. For purposes of geographic analysis, the same map would be improved tremendously if it were to include quantitative information indicating "how much" was produced at each plant by the assignment of appropriate numerical values to each dot or to each dot of a different size. Such a map shows two related data and illustrates important, although general, information about the aluminum industry. If the researcher were to refine the map further by plotting the sources and amount of bauxite (aluminum ore) and by adding isolines indicating power costs, it would be discovered that the bauxite sources and the aluminum plants have little relationship spatially, but that power costs appear to be most significant since all aluminum plants are located in very low power cost areas. Thus, a relationship has been unearthed, one which may be further investigated and tested by using statistical techniques discussed later in this chapter.

Ratio maps are usually significant to geographic studies. They require previous computation of data as they show relationships of amount to area, percentage of one or more items to the total, or some other statistical relationship between two or more phenomena. Simple examples are the widely-used population density maps (*number* of persons per given unit of *area*), or effective growing season maps (80 *percent* of the *average* frost-free season).

Some ratio maps may be quite complex. As a hypothetical example, let us assume that in a given subhumid area, the amount of moisture is a critical factor in determining whether or not it will be economically feasible to attempt to grow a given crop. Let us also assume that this crop needs 9 inches of surplus rainfall (beyond the evaporation rate) during the growing season (May-September), and that the area receives 8 to 10 inches depending upon the locality. Let us assume further that data indicating total monthly rainfall and evaporation are available for various locations. The question might be, "What are the chances of success in any given year at locations A, B, C?" This may be shown by constructing a map with several isolines, each representing a given percent of years in which 9 inches or more of surplus water may be expected. Such a map requires several manipulations of data: the average monthly evaporation rates subtracted from the average monthly rainfall rates for each location during the growing season, computing the percentage of years that each location will receive at least 9 inches of surplus water during the growing season, and plotting several isolines (perhaps the 80 percent, 90 percent, etc.) on the map.

The researcher faces two pitfalls in constructing cartographic devices for analysis. There has never been a map constructed of the world, or of a large segment of the earth, that is not distorted in some degree. It is impossible to represent a curved surface accurately on a flat plane; and small scale projections may show either true shape or true area or, perhaps, neither. It is not likely that the student researcher who is engaged in his first research problem will be dealing with a matrix so large that the earth's curvature will be significant. However, in the event that he does deal with spatial distributions on a world or a continental scale, care must be taken to select the appropriate projection that illustrates his data without distortion. The second danger is inherent to all map construction. Values must be assigned for any dots, lines, or legends which are used. Extreme care must be taken and a well-defined rationale stated for assigning numerical values or legend classification to ascertain that relevant spatial patterns are not obscured or that actual misrepresentation results.

To illustrate this point with very simple examples, let us assume that a researcher is constructing a contour map of an area (lines connecting points of equal elevations) and that he decides to use 20-foot contour intervals. If all the area is under 20 feet elevation except for one small hill summit, the map will show only the outline of this one small area, and the rest will appear to be flat. In reality, the area may be a "bad lands" with nearly perpendicular slopes and sharp gullies between 0 and 19 feet in elevation.

Let us assume further that a researcher constructs a map of an area to illustrate the percentage of the population that has blue eyes. He elects to show this spatial distribution by devising a legend with categories such as 90 percent and above, 70 to 90 percent, and so forth. Perhaps the map might show a small island in the 90 percent + category surrounded by a very large area in the 70-90 percent category. In the latter area, perhaps 88.7 percent of the population is blue-eyed. By modifying his legend values downward by only two percentage points, the researcher has changed his spatial distribution drastically. A general rule to follow in assigning values to legends, isolines, dots, and the like is to ask the question, "Will the proposed value categories show the spatial distribution of the significant aspects of the data as they actually exist?"

Statistical Analysis and Presentation of Data

In data presentation, the most used expressions of measurement are those of descriptive statistics, such as the mean, median, mode, standard deviation, locational quotient, variance, quartiles, and similar tools. Bar graphs, flow charts, scattergrams, pie charts, and a number of related tools may be developed using statistical procedures, but may be presented cartographically. In organizing and analyzing data, the importance lies in the accuracy of the operation and its appropriateness for the task, not in whether it is part of a particular technique.

Special techniques of spatial analysis involving more sophisticated uses of statistical procedures have been developed recently. This development has been spurred by new techniques of inferential statistics, by the creation and widespread use of the electronic computer, and by the evolution of geography into a more exact scientific discipline. Since science seeks to be as precise as possible, geography has accepted the need for more exact measurements. This need is as great for the normative sciences as it is for the experimental sciences. Numerical statements are more precise than verbal expressions. They are also more universally understood.

It is well to remember that geography is not an experimental science. Geographic analysis does not contemplate the effect of controlled variables, but it observes the locational pattern as it is found and attempts to account for its distribution. Since the number of independent variables effecting any distribution is large and not experimentally controllable, it is necessary to accept the philosophy of *improving* the accuracy of the analysis as a major goal of research geography. Total explanation may be ideally desired, but often it is impossible, or at least too time-consuming and expensive to be feasible. The large number of uncontrollable variables makes it advisable to employ some statistical method of identifying the effect of the variables under consideration. Thus, their separate effect on the distribution, as well as the total amount of explanation accomplished by the study, can be measured with some accuracy.

At the present time, the average geography student has not been exposed to statistical procedures to the extent that he has been trained in verbal expression. It has been only recently that departments of geography have begun to require courses in quantitative techniques. It may be assumed that, as the student advances in his professional training, he will be exposed to increasingly more sophisticated statistical techniques. However, for the beginning researcher, a working knowledge of a few statistical techniques will, with the help of a basic statistical manual, make it possible for him to understand articles in geographic publications and to engage in research studies.

The research student can perform most of the elementary associative analysis required and read most of the published quantitative findings in geography if he understands three basic statistical procedures. In the geographic literature, there are approximately ten statistical research tools that have been found to be useful. In all the articles published by the major geographical journals between 1954 and 1966, plus the *Regional Science*

Journal (1959-1965) and *Papers* (1955-1965), only nine separate statistical tools were used to any degree. Out of a total of 172 separate uses of these tools, regression-correlation analysis was used 104 times.[1] It would appear that knowledge of regression-correlation, plus some elementary model concepts, would equip the undergraduate and the beginning graduate to research and analyze a major part of most spatial problems. The full extent of his specific needs can be determined only by the individual and his advisor in the light of the research problem faced. The skill of analysis cannot be measured in terms of the number of techniques and tools used. The purpose of the research is to answer the questions asked. The selection and use of the proper procedures are only part of the problem-solving process. With this in mind, let us consider now the three most commonly used statistical devices in geographic research: the scatter diagram (scattergram), regression analysis, and the coefficients of correlation.

THE SCATTER DIAGRAM

The relationship among areally associated phenomena is usually not apparent until some device for detection is applied to the data under observation. One such device is the scatter diagram, or scattergram. It is easy to construct, uncomplicated in nature, and easily interpreted. It not only visually demonstrates the general character of the relationships, but it directs attention to the cases which do not conform to the pattern.

The scattergram is a device for visually showing the relationship between two variables for a selected number of observation units. One variable is measured on a horizontal axis and the other on a vertical axis. Where the two lines intersect, a dot is made indicating the location of that observation unit on the diagram. For example, the average maximum summer temperature (June, July, and August) for Phoenix, Arizona, is 102° F., and the elevation is 1000 feet. Phoenix is represented on a scattergram as a point, as shown in Figure 5. Yuma at 104° F. and 138 feet in elevation is located higher on the temperature axis but lower (to the left of Phoenix) on the elevation axis. Tucson at 2410 feet above sea level is to the right of Phoenix on the "X" axis, and its summer maximum temperature (99°F.) is lower on the vertical scale.

1. Harold L. McConnell, "Statistical Methods in Contemporary American Geography: A Bibliography from Selected Recent Journals," 1966, mimeographed. Iowa City: University of Iowa.

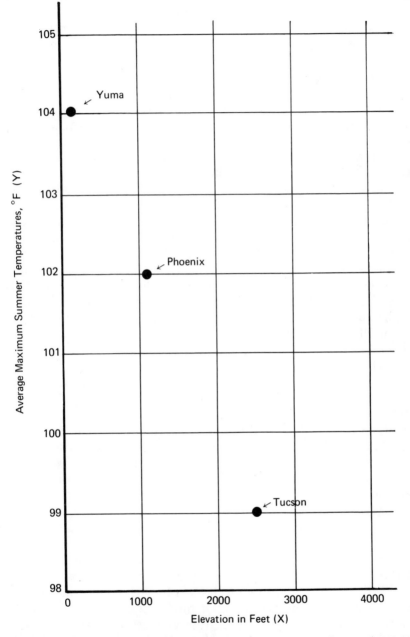

Figure 5. Scatter diagram showing average maximum summer temperatures and elevations for selected stations in Arizona.

There are three questions of importance to note in reading a scattergram: (1) Do the points tend to be arranged in a sloping straight line? (2) Does the line slope from upper left to lower right, or from lower left to upper right? (3) Do the dots arrange themselves close to the line of slope? Each of the answers reveals something about the nature of the data and directs future research. The three questions should be considered separately.

If the dots are arranged in a sloping line, a relationship between the two variables is indicated. If the dots are widely scattered, are circular in arrangement, or are in a straight horizontal or vertical line, no relationship exists.

Since the scale of the vertical axis (Y) usually progresses from bottom to top, and the horizontal scale (X) runs from left to right, the slope of the line gives additional information. If it slopes from lower left to upper right, this means that, as the magnitude of one value increases, so does the other. This is called a *positive or direct* correlation. However, if the line slopes downward from the upper left corner to the lower right, it shows a *negative or indirect* relationship between the variables. This is the case in Figure 5, where Yuma with a low elevation has a high temperature, while Phoenix and Tucson have higher elevations with lower temperatures. Of course, any scattergram used in a research project will have more than three observation units; but those in Figure 5 illustrate the inverse correlation.

The third characteristic to be noted is the "closeness of fit," that is, how close the dots are to a straight line. As mentioned previously, a scattering of dots shows no relationship. The nearer the dots tend to be to a straight line, the higher the correlation between the variables. The exact degree of this "fit" is called a coefficient of correlation. It may range from no relation (zero) to a perfect or 1.00 correlation; and it may be either positive ($+$) or negative ($-$), depending on whether the slope tends upward or downward.

Another aspect of a "close fit" should be mentioned at this time. Correlation statistics are based on the assumption of a linear relationship between the variables. If the dots tend to follow a straight line with little scatter, the researcher has good reason to assume that there is a linear relationship.

Probably the main research use of the scattergram is during the search for hypotheses which might explain the relationship between variables. The scattergram gives the researcher a quick idea of the degree of relationship. If the observations cluster along a sloping line, further investigation is suggested. At the same time, the scattergram calls attention to the isolated case (residual). The residual may become the subject of a separate "case study." More often, its value is to direct attention to the presence of a strong independent variable which, when considered in the analysis, will not only explain the deviation of the residual but will improve the fit of all other observation units in the study.

For example, let us assume a hypothetical case where the researcher attempts to develop a system for classifying villages between 3,000 and 5,000 population. He hopes to classify them on the basis of occupational structure; and he has theorized that manufacturing and selected service occupations (persons employed in grocery stores, service stations, barbershops, etc.) will prove meaningful. He plots on the scattergram the percentages of the total population employed in manufacturing and service occupations for each of the villages in his research area (see Figure 6). The clustering of dots in distinct groups indicates that there is

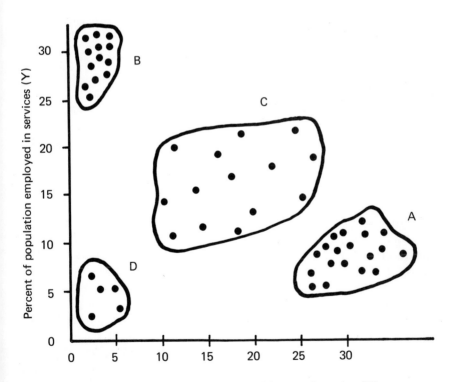

Percent of population employed in manufacturing (X)

Figure 6. Scattergram constructed for classifying villages of 3,000 to 5,000 population on the basis of occupational structure.

some basis for his proposed classification. The information brought to light by the scatter diagram enables him to classify villages tentatively as follows:

Group A—"Manufacturing Villages," which he tentatively defines as those with 25 percent or more of the population employed in manufacturing and no more than 15 percent in service occupations. Perhaps this confirms an earlier hypothesis that villages of this type are located near large manufacturing plants on the periphery of large cities.

Group B—"Service Centers," which serve as central places for adjacent rural areas (tentative definitions can be stated as in Group A villages). Perhaps this confirms a hypothesis that villages of this type are located some distance from larger towns or cities.

Group C—"Diversified Villages," which have a wider range of functions and which may represent a broader or balanced occupational structure.

Group D—"Strange Cases," where manufacturing and service occupations are relatively unimportant.

The researcher now finds that the "Strange Cases" require further thought, and he hypothesizes that these may be farming communities, mining towns, or settlements inhabited by persons employed in commercial fishing or in some other occupations that he has not considered. Perhaps they may even represent a resort or a retirement community. He must now investigate further if he wishes to subdivide the "Strange Cases" grouping into meaningful categories.

This hypothetical example illustrates a way in which scatter diagrams may be used to analyze data. In this case, *one* diagram has provided the researcher with information that may serve as the basic framework to classify villages, and it also focuses his future efforts toward refining his classification (perhaps percentage of gainfully employed rather than total population would be a possibility). The scattergram also directs the search for other variables to understand and explain the "no fit" cases.

In the event that such a clustering of residuals is found to be concentrated within a limited area, there is sound basis for defining a new region or for recognizing a subregion. For example, northern Missouri has a section referred to as "Little Dixie" which posesses unique voting characteristics. In the regionalizing of political patterns, this section commonly appears as a distinctive subregion. In accounting for the area, the independent variable of southeast American immigrants and their cultural views would have to be considered.

REGRESSION ANALYSIS

A regression analysis is one method of measuring the relationship between two or more variables. If only two variables are considered, it is called simple regression as contrasted with multiple regression when three or more variables are considered. Regression analysis is a precise mathematical expression of the observation units plotted on a scattergram. It defines the direction and degree of slope of a line which is drawn closest to the average of all the data on the scattergram. This line is called the "line of best fit" or the "regression line" and is designated here by the symbols "Yc." The equation for "Yc" is $Yc = a + bX$ for simple regression. In the case of multiple analysis, additional "bX" values are designated, as $Yc = a + bX_1 + bX_2 + \ldots bX_i$. Discussion here will be limited to simple regression, but the same procedure may be applied to multiple regression analysis.

Two characteristics of the regression line are necessary for its description. First is the elevation of the line as established by where it is started on the vertical (Y) axis. This value is called the "Y intercept" and is designated by a small "a" (see Figure 7). The second characteristic is the slope of the line or coefficient of regression, designated as "b." ("X" is the magnitude of the variable measured on the horizontal axis.) The actual procedure for finding the value of "a" and "b" is discussed at the end of correlation methods in Appendix F. Note that the two characteristics are independent of each other, but that both are needed to establish the position of the regression line.

The "a" value does not have particular significance in an individual analysis beyond giving the line its starting point. It may, however, be of great value in future studies which use the regression analysis as resource material.

The regression coefficient ("b") shows the relationship between the variables. It expresses the change in the "Y" variable as it varies in relation to the "X" variable. The "b" value answers precisely some of the questions previously raised concerning scattergrams. The direction of slope is indicated by the sign preceding "b," and the incline of the slope is indicated by the number. A minus sign means that an inverse relationship exists, with a downward sloping line signifying this.

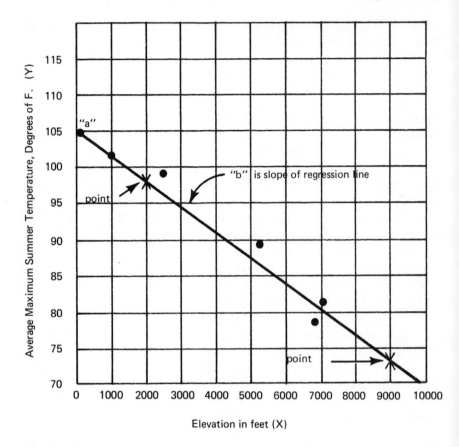

Figure 7. Elevations and average summer maximum temperatures for six selected stations in Arizona. Source of data: TABLE 3.

One question cannot be answered by "a" or "b." That is, how close are the observations to the regression line? Such a measure of the variation of the dots from the regression line may be obtained by computing the standard error of the estimate. It simply indicates how much of the variation of "Y" is not explained by the changing value of "X." It is computed by

$$S = \sqrt{\left(\frac{\text{Sum of Squared Variation from Regression}}{\text{Number of Observations} - 2}\right)}$$

The standard error of the estimate indicates, in absolute terms, on the average how much the dependent observations are to be expected to vary from the regression line. Thus the geographer, knowing his analysis is not perfect, is informed of the amount of variation to be normally expected. Armed with this knowledge, he can ignore cases which are probably insignificant variations and can direct his analysis to truly exceptional cases when they occur. A more useful expression of the closeness of fit may be found in the coefficient of determination. This is an extremely valuable tool for analyzing spatial relationships and will be discussed in detail later in this chapter.

The method of fitting a regression line to a distribution of matched variables when "a" and "b" are known is demonstrated by the following example using actual temperature and elevation data for six locations. The elevation in feet of the locations is designated as "X," and the average summer maximum temperature in degrees of Fahrenheit as "Y." The first step is to prepare a table of values (see Table 3).

TABLE 3
Elevations and Temperatures for Selected Stations in Arizona

Station	(Elevation) X	(Degrees F.) Y
1	6903	78
2	1083	102
3	5219	89
4	6964	81
5	2410	99
6	138	104
	$\overline{X} = 3786$	$\overline{Y} = 92$

From the table, the researcher may construct a scattergram, plotting the data in order to analyze the distributions better (see Figure 7). This procedure, although not absolutely necessary, is advisable in order to fit a line of regression to the data. (In multiple regression problems, it is not possible to show the variables on a scattergram.)

The regression line, or line of best fit, is drawn on the basis of all paired observation units and is the closest straight line to each and every dot representing an observation unit. The determination of this line is demonstrated in Appendix F. Using the data in Figure 7, the line is based on a = 105 and b = –3.52X. It is drawn on the diagram by selecting two points on the "X" axis, multiplying the value of each by –3.52, and adding the product to 105. Both of these points are then plotted on the scattergram, and a straight line is drawn which passes through the points. A good test of its accuracy is to determine if it also passes through the mean of "X" and "Y." If it does not, some error has been made.

The regression line's relationship to temperature variations is more meaningful than an average line ($\overline{Y}$) which, in this case, is 92° (it would be shown as a straight horizontal line). An estimate of the temperature of Phoenix on the basis of the average would be 10°F. in error. By using the elevation variable in a regression analysis, a temperature of 101.2°F. (a + bX) would be predicted. Since Phoenix actually has 102°F., the analysis has failed to explain 0.8°; but it has "explained" 9.2° of the error as estimated by the average. It is obvious that a great deal of the temperature variation from the average can be accounted for by the variation in the elevation of the station.

The distance of the observation dot from the regression line compared to its distance from 92° ($\overline{Y}$) indicates the amount of "Y" variation explained by regression. Some stations are not completely explained by elevation, and some are over-explained. Other variables besides elevation (such as local atmospheric conditions, latitude, and exposure to air masses) affect the temperature, and a more complete explanation could be obtained by a multiple regression analysis. The need for selecting additional variables depends on the nature of the specific problem.

COEFFICIENTS OF CORRELATION AND DETERMINATION

Two useful and frequently encountered statistics in geographic analysis are the coefficients of determination ("r^2") and its square root ("r"), the coefficient of correlation. In the preceding discussion of regression analysis, the relationship of variables was described as to direction and slope of the regression line by the formula, Yc = a + bX. Furthermore, it revealed that some measure of scatter, that is, how much error is contained in the analysis, could be stated by the "standard error of the estimate" statistic. A more useful measure is the coefficient of correlation and its related statistic, the coefficient of determination.

The coefficient of determination ("r^2") is a measure of the relative relationship between the X and Y variables. In the regression analysis of temperature and elevation (Figure 7), a certain amount of unexplained variation in the distribution of temperature (Y) remained after the regression analysis. Theoretically, all spatial variation could be accounted for, but in actual research, geographers rarely provide complete or total explanation. The coefficient of determination (r^2) is a measure of what percent of the total variation in "Y" is explained statistically (or determined) by the distribution of "X." Referring again to Figure 7, Phoenix with its 102°F. average summer maximum temperature is 10 degrees above the average of 92°F. Regression analysis accounted for 9.2°F. of that variation, or 92 percent of the total variation of the Phoenix temperature from the average. In this single case, the coefficient of determination is .92, i.e., $r^2 = .92$. Of course, no researcher would cite an "r^2" for one observation because of the unreliability of such a figure. But this does illustrate how the coefficient of determination is a measure of how much of the variation is explained by the regression analysis.

In Figure 7, every observation is imperfectly explained. In other words, the line does not pass exactly through any dot on the diagram. The equation clearly explains only part of the total, but how much? In this case, an "r^2" for all the observations is .994, which means that for these six stations in Arizona, elevation accounts for 99.4 percent of all variations in temperature. Other variables must explain the remaining 0.006. If some other significant variable were combined in the analysis to arrive at a multiple coefficient of determination (R^2), the resulting figure would range somewhere above 0.994; but it is unlikely that it would ever reach 1.00 or 100 percent explanation of the distribution.

There are various statistical correlations for arriving at coefficients of determination. Some are more precise than others, but all essentially define covariation. For the purposes of the beginning researcher, three of these correlations are discussed in the order of their ease of computation. In the cases of these three correlations, the more easily the coefficient can be computed, the less precise is its measurement.

Tetrachoric Correlation

The tetrachoric correlation is the easiest to compute. It is useful for rapid calculations in preliminary research and for class report preparation. It is also occasionally found in published research reports. It does lack precision, but its results are usually close to those of more complex methods. The correlation is computed by ranking the variables and counting the number of observations which are in the top half of the array (above the median) for both variables. This number is then divided by the total number of observations to obtain a percentage value. For example, if ten observations were ranked from 10 to 1 in value, and four of them were in the top half of *both* "X" and "Y" (see Table 4), the value would be 40 percent. This value corresponds to a coefficient of correlation (r) of 0.81 and may be read directly from a table of percentage values and correlation equivalents in Appendix F.

TABLE 4

Hypothetical Data Illustrating a Tetrachoric Correlation

Obser- vation	Rank of "X" Variable	Rank of "Y" Variable	
A	10	10	
B	9	9	40 percent of both
C	8	8	"X" and "Y" rank
D	7	7	above the median.
E	6	1	Median Read Coefficient of
F	5	6	Correlation in
G	4	5	Appendix D.
H	3	4	
I	2	3	
J	1	2	

The temperature and elevation data from Table 3 may be used to demonstrate the computation of a negative tetrachoric correlation:

TABLE 5

Elevations and Temperatures for Selected Stations in Arizona Illustrating a Negative Tetrachoric Correlation

Station	Elevation "X"	Rank of "X" Variable	Temperature "Y"	Rank of "Y" Variable
1	6903	5	78	1
2	1083	2	102	5
3	5219	4	89	3
4	6964	6	81	2
5	2410	3	99	4
6	138	1	104	6

In Table 5, there are no observations which are in the top half of both "X" and "Y." The percentage, therefore, would be zero, and the coefficient of correlation by this method is –1.00, a perfect negative correlation. The coefficient of determination (r^2) is 1.00. The tetrachoric correlation indicates perfect prediction, which is not correct as may be seen by the scattergram in Figure 7. A more precise method of correlation will give a more accurate correlation value.

Rank-Order Correlation

A second method of correlation is the rank-order correlation. It is also simple to compute and is frequently used. The formula for its operation is:

$$1 - \frac{6 \ \Sigma \ d^2}{N^3 - N}$$

where "N" is the number of observations, and " Σd^2 " is the sum of the squared deviations. Again using the temperature-elevation data, the correlation is obtained as follows:

TABLE 6
Elevations and Temperatures in Rank Order
for Selected Stations in Arizona

Station	"X" Rank	"Y" Rank	d	d²
1	2	6	4	16
2	5	2	3	9
3	3	4	1	1
4	1	5	4	16
5	4	3	1	1
6	6	1	5	25
			$\Sigma =$	68

$$1 - \frac{(6)(68)}{(6 \cdot 6 \cdot 6) - 6} = 1 - \frac{408}{210} = 1 - 1.94 = -.94$$

$$r = -.94 \qquad r^2 = 0.884$$

By the rank-order method of correlation, the portion of the distribution of temperatures explained by elevation is 0.884. This differs somewhat from, and is a more precise measure than, the 1.00 coefficient obtained by the tetrachoric method.

In order to gain experience in the use of rank-order correlation, let us consider another problem which tests the hypothesis that elevation affects temperature range:

TABLE 7
Elevations and Temperatures for
Selected Stations in Europe

Station	(Meters) Elevation	(°C) Temp. Range	"X"	"Y"	d	d²
Bozen	290	22.5	1	7	6	36
Buxen	580	21.9	2	6	4	16
Innsbruck	600	21.1	3	5	2	4
Sterzing	1000	20.7	4	4	0	0
Schafberg	1780	14.6	5	2	3	9
St. Bernard	2470	15.2	6	3	3	9
Sonnbuck	3105	14.2	7	1	6	36
					$\Sigma =$	110

$$1 - \frac{(6)(110)}{(7 \cdot 7 \cdot 7) - 7} = 1 - \frac{660}{336} = 1 - 1.97 = -.97$$

$$r = -.97 \qquad r^2 = 0.94$$

The correlation indicates that .94 of the variation is explained by elevation.

Pearson Product Moment Correlation

The method used to measure most accurately the coefficient of determination at 0.994 for the six weather stations in Table 3 is the Pearson Product Moment Coefficient of Correlation. This is the most widely used and best known correlation tool in geographic research. Its computation is comparatively slow and requires careful preparation. As a result, it is somewhat difficult to use. Consideration here of a few points regarding its use and value should render it usable to the beginning student.

The Pearson Product Moment Correlation method is so commonly used that almost every academic institution has it already programmed for the electronic computer. These "call programs" have been designed by the computer manufacturers, and the researcher has only to supply the data in the proper form. The computer is especially needed for the working of multiple correlations, where more than one independent variable is used.

When only one independent variable is used, the process is called simple correlation. Its computation is simple when compared to multiple correlation. However, even this computation is difficult to perform unless careful attention is given to each step. The beginning research student will want to master the simple correlations for several reasons. First, since two variables can be shown on a scattergram, the student researcher can actually see the unaccountable variation by the scatter of the dots from a line. He may compare this visual image with the measurement of correlation and develop an understanding of what a correlation means. Second, simple correlations can usually be computed more rapidly by hand than by computer because most computers are too busy for the student to gain quick access to their facilities. In fact, some computer centers consider simple regression problems too easy to justify their use unless a great number of variables is involved. Finally, the method is not so difficult that it cannot be used by any careful worker who is able to add, subtract, multiply, and divide without error.

In order to demonstrate how the Pearson Product Moment Coefficient of Correlation is computed, a simple problem of spatial relationship is worked step by step in Appendix F.

The interpretation of a particular coefficient of correlation is more useful when it is subjected to a test for significance (see Appendix F). There are various tests for significance which tell the researcher that the chance that a coefficient as high as the value he obtained could be obtained by the accident of sampling. Among such tests, the Chi Square, "t," and "F" are most common. If a value is significant at the .01 level of confidence, it is usually considered "highly significant" because the odds are only one in one hundred that a relationship as high as the one found could be obtained by accident. Another commonly cited level of confidence is the .05 level, referred to as "significant."

While the significance of any particular coefficient of correlation is directly related to the number of observations in the sample considered, some general values may be placed on various coefficient levels. In general, coefficients below .25 are not indicative of a meaningful relationship between or among variables. Between .25 and .50 suggests a small to moderate relationship, while .50 to .75 may be considered substantial. Above .75 is considered high, or very high if it is over .90.[2]

MODELS

The construction and use of models by the researcher has long been an important aspect of geographic research. Some cartographic models, such as aerial photographs (iconic models) and maps (analog models), have altered the scale or substituted a property in order to study reality better. In recent years, emphasis has been directed toward the use of symbolic models, especially mathematical symbols to define the parts of the model, and graphic models of various types, especially those dealing with space-function relationships.

The beginning researcher should know how to construct and use symbolic as well as more standard types of models in his work. A symbolic model is a description of a set of phenomena stated in a way that allows any change in the various parts to be measured and its effect on the whole to be predicted. These models focus on a set of theoretical phenomena and may or may not reflect reality. Additional variables, however, may be added to a model until reality is approached; but the basic model is designed to consider the interrelationship of a selected set of variables. Consider, for example, one of the simpler models—the regression equation for predicting temperature change. If the average lapse rate of $3.5°F$ per 1,000 feet is expressed as $[-3.5 \ F \ (X)]$, then any change in either the elevation or the rate of temperature variation can be assessed. It is evident that, in reality, many other variables affect temperature change. A nonscientific thinker whose knowledge of the "reality" of temperature change is limited to the use of the home refrigerator might depreciate the value of this model which predicts temperature decrease in relation to increased ele-

2. J. P. Guilford, *Fundamental Statistics in Psychology and Education* (New York: McGraw-Hill Book Co., 1965), p. 145.

vation. However, for the climatologist and the geographer, such a predictive model has proved to be the basis of a major concept concerning atmospheric processes.

Because of the many types and subtypes of models, no attempt at a comprehensive discussion is made here. A few examples will serve to illustrate their usefulness. A graphic space model depicting the incidence of serious crimes for a major city as they are spatially related to the home of the criminal has been constructed (see Figure 8). This is a single variable space model, and its use is mainly as an hypothesis to serve as a way to observe the distance-function elements of a spatial problem. In itself, the space model is not intended to be a solution to a problem. Such models may become extremely complicated as additional variables with varying degrees of effect are added. It is apparent that this complication would soon limit its usefulness and that some form of mathematical definition of the parts would be more manageable.

Another example of one of the more common mathematical models is the gravity, or interaction, model mentioned in Chapter 2. Based on the law of gravity, such models define spatial interaction. The most elementary type treats only two variables—population and distance. The interaction between the points is predicted as equal to the product of the population of the two cities divided by the distance between them:

$$ i = \frac{P_1 \ P_2}{d} $$

The resulting figure is an index of the interaction which may be compared with a similar index figure for other cities. To illustrate, let us consider a model based on the following data:

TABLE 8

Populations and Distances between Four Hypothetical Cities

City	Population (000)	Distance in (00)			
		Metropole	Suburbia	Urbann	Bigtown
Metropole	10	0	4	6	7
Suburbia	20	4	0	5	8
Urbann	30	6	5	0	1
Bigtown	40	7	8	1	0

If there were 100,000 automobiles traveling between Metropole and Urbann, how many would be expected to travel between Suburbia and Bigtown? Using the mathematical interaction model, the following results are obtained:

Case	Cities	Population X Population (000)		= Total	÷ Distance	= Index
A	Metropole Urbann	10	30	300	6	50
B	Suburbia Bigtown	20	40	800	8	100

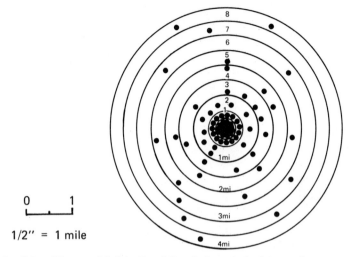

0   1

1/2" = 1 mile

Figure 8. Crime-Distance Model. Domicile of the criminal is at the center of the diagram. Each dot indicates where one percent of the total crime is committed in relation to the criminal's domicile. Lines denoting zones are representative of one-half mile intervals.

It is readily apparent from the computation that for every 100 automobiles traveling between Suburbia and Bigtown there are only 50 cars traveling between Metropole and Urbann. This is a ratio of 2:1. Therefore, 200,000 autos would be expected to travel between Suburbia and Bigtown, since 100,000 traverse the Metropole-Urbann route.

As different phenomena are considered, modifications must be made in the model to better define the interaction. For instance, the interaction rate of airplanes between two points would differ from ground retail trade traffic. Therefore, constants and other variations in the formulae are needed to better predict the interaction. Such models can predict only on the basis of the variables employed. In reality, the element of accident, or chance, may partially defeat the most carefully drawn conclusions. Fortunately, if more "reality" is desired in model prediction, there are models which consider the element of chance. These are called "stochastic" models. They add an ingredient of randomness to the structure of the model. This is a partial answer to the criticism that models do not sufficiently resemble reality. By incorporation of the chance factor, the stochastic model predicts what *might* happen and what the probability of its occurring is.

Conclusions and Prediction

The final stage in a research problem is to state conclusions based on the analysis of acquired data. All conclusions must be either carefully documented by utilizing pertinent specific data or, if based on the opinion, or judgment, of the researcher, they must be stated as such and the rationale, or logic, for these opinions clearly explained. It is important to realize that, if the research question cannot be answered completely or the hypotheses confirmed, even though the research work was carried out in a sound and scientific manner, the results of the study are still valuable. In research work, it is not possible to answer the question or to come to positive conclusions in all cases. In the final analysis, the success of a research problem is judged on the manner in which it was carried out rather than whether the conclusions were positive or negative.

If the research question has been answered and the hypotheses confirmed, either fully or partially, the researcher can then project his findings and predict outcomes in situations similar to his research problem and matrix. A desired result of research is this ability to predict future situations. The researcher will find great satisfaction in his research work if the conclusions contain some element of prediction. Such prediction is also a purpose of science, although every scientific study does not have to be predictive in nature. It is further true that such prediction is usually limited to the time and place of the universe of the study. In the long range of science, universal laws are built from the bricks of limited predictions. Any time a norm, a law, or a history is established, a certain amount of prediction is possible for the universe studied. Their predictive value in a larger or otherwise different universe must be determined separately.

One of the best methods for the geographer to utilize for predicting is to construct a physical or conceptual model of an object or situation. This model becomes the basis for comparison, and deviations from it are apparent and measureable. For example, if the model of a city has a major business district near its center, any variation from this norm would attract attention. Geographers have devoted considerable time to constructing spatial models and adapting models from other sciences to meet spatial needs. An example of the latter case is the gravity, or interaction, model discussed previously. Obviously, the model is based on the law of physical gravitation. Such models, when properly modified to meet varying situations, are quite helpful in spatial prediction.

The number and kinds of models, analogies, norms, and other predictive devices are so great that it would not be practical to catalog them in an introductory book to geographic research.

chapter 6

writing
geographic research reports

Research which is unrecorded, or inadequately recorded, has little value to anyone but the researcher. Undoubtedly, much research as not been made available to others because the study has not been written, or because it has been written in an unsuitable form. Writing a geographic research report is often difficult for the beginning research student, but it is an important part of the research project. It must be approached with the same sense of scholarship as the researcher feels toward the other components of the research problem. It requires skill, organization, insight, and hard labor. A part of this task is the proper arrangement of the physical materials needed in order that the mind may be free to concentrate on the subject matter.

The Work Area

The fact that writing involves a considerable amount of creativity is all the more reason that the tools and necessary equipment should be assembled in advance and be readily accessible. This necessitates having all tools and supplies in an established location. It is not as important that there be a particular arrangement as it is that the equipment always be in the same location. However, as the student geographer should know by this time, organization of space is important at the microscale as well as at the macroscale.

The writing of the geographic research paper requires certain basic supplies. Some of these may be used up rapidly, and so they should be available in sufficient amounts so as to ensure that there be no need for leaving the work desk to obtain more. These supplies include pencils, pens, a ruler with both metric and English scale, and sufficient paper. It is advisable to have some extra note cards with the supplies.

It is likely that the writer will need extra base maps and colored pencils for making preliminary maps. Frequently, ideas occur during the writing process, and it behooves the writer to record his thoughts immediately. Existing maps should be included among the working supplies. There should also be available source books, such as a good dictionary, a thesaurus for word choice, a grammar for form and structure, atlases pertinent to the subject, a cartographic manual, special writing or style manuals determined by the school or by the ultimate authority who will judge the work, and basic technique materials, such as quantitative tables. In short, any reference book or source material to which the researcher may refer frequently should be readily accessible.

Preparing to Write

With notes sorted and supplies and sources gathered, the student is ready to go into seclusion preparatory to the actual writing. Scholarly writing does not come easily to most people. One of the rules for the writer to remember is that his first task is to communicate with the reader. No matter what else he has done or will do, it will be lost if he does not honestly communicate. This includes the use of clear, simple sentences, and the avoidance of jargon. He should always remain objective, write in the third person, and avoid the editorial "we." This does not imply stilted or awkward usage, whether writing in the active or in the passive voice. Since the study has already been conducted, the past tense is mainly used throughout the report; but the present and future tenses may be used where appropriate.

It is not by accident that the phrase "a scholar and a gentleman" has been so widely circulated. The scholar is a gentleman in the sense that he is helpful, courteous, and considerate. If an error is found in the work of another, it is not introduced into the research report unless it has a bearing on the problem. If it is a part of the study, the original source and the accurate correction should be given. Modesty is a virtue highly recommended, especially for beginning scholars. Albaugh states that an overly positive, dictatorial air, inappropriate in any scholarly writing is unforgiveable in a graduate thesis.[1] The same would be true of research reports in general. The scholar is definite, and he is precise. He does have opinions and states them; but while he is positive, he should use due caution. If a conclusion appears to be justified from the evidence, he would probably say "it appears that" rather than "it is obvious." When the opinion is not his own, the scholar is careful to give credit to the source.

The First Draft

The writing of the first draft of a geographic research report can be an extremely pleasant period for the researcher. He is at the stage where, like the artist at his easel, he sees the planning and work of his project take visual form. The pleasure he derives from his own creation can be a great reward for any scholar. However, the writer should prepare to give the composition his full attention as interruptions at this point may be very destructive to his creative capacity.

The first or rough draft is intended to be the framework within which the researcher organizes and records his findings and ideas in a systematic form. It is the rough sketch of the final creation and is not intended to be widely circulated. Consequently, it should be written as rapidly as is consistent with good thought. It is hoped that style and usage are innate capabilities with the writer since he should not devote extensive time to such matters while composing the first draft. If mistakes in grammar or style occur, they may be noted; but detailed editing should not be done at this time. All writing should be double or triple-spaced in order that inevitable corrections, additions, or modifications may be made later. It is desirable that only one paragraph be written to a page so that the order of thoughts may be rearranged later. Quote note cards and other notes may be stapled directly to the appropriate manuscript page for convenience. Footnotes are essential, but a short form should be used because little time should be taken in making such notations. The name, year, and page should certainly suffice here.

Exactly how the written draft is handled varies with the length and type of geographic research report. A term paper may be submitted to the instructor for tentative approval, and a graduate thesis may be submitted a chapter at a time. In any case, the

1. Ralph M. Albaugh, *Thesis Writing* (Totowa, N. J.: Littlefield, Adams & Co., 1962), p. 11.

first draft should be set aside to "cool" for a period of time. After an interval, it is ready for critical review; and the task of the second writing begins.

The Second Draft

The term "second draft" is really a euphemism for a considerable amount of writing that is necessary between the first and final draft of the geographic report. While it is common to think of writing three drafts of the research paper, one should realize that many more may be necessary. Seasoned writers frequently write a page ten, or more times before it is satisfactory. Certainly, the beginning writer should not be disappointed if his second writing is not satisfactory in all respects. The second draft is a refining process that advances the manuscript from the rough recording of facts and ideas to the stage where the various components of the report mesh and can be viewed in perspective.

It is here that the writer works on cohesion of sentences and paragraphs. Reorganization often is necessary so that the sequence of thoughts is presented in the most logical manner. Errors of detail which were allowed in the rough draft are corrected, and unclear points may need elaboration and clarification. Verbosity and clichés of thought or expression are pared from the manuscript.

In a geographic report, the rough draft includes illustrative materials (sketched hastily or made with pencils) which need correcting and revision. These include maps, charts, and diagrams used or referred to in the first draft. Like the painter who is filling in the color and correcting the format from his preliminary sketch, the research writer now sees his creation taking the form which gives pleasure to him, the scientist—a research report which communicates from writer to reader with simplicity and power.

The Final Draft

The final draft would be anticlimactic were it not the culmination of the entire project, the end toward which the research project was aimed. Even so, it is primarily the final editing of the completed report, the refining, perfecting, and clearing of miscellany.

In the final drafting, an attempt is made to obtain perfect sentence structure and spelling. Quotes and footnotes are given their last scrutiny and approval. The final bibliography is assembled, and all maps are reproduced for inclusion in the completed report. Since this is the final form of the report, page numbering, indexing, making content tables, and all similar items must be completed at this time. When the final draft is completed, the manuscript is ready to go to the typist or typesetter, as the case may be, to be put into its final form.

The Report Content

The general content of the scientific research report is essentially the same regardless of its magnitude or purpose. When the research steps described in previous chapters have been carried out, the purpose of the written report is to record accurately the findings and results of the project. The scope and length of a given report will vary, depending upon the research problem and its *raison d'être*. Normally, term papers, class projects, and seminar papers are concerned with problems smaller in scope than are those for a Master's thesis. The Master's thesis is probably the most universal requirement in American colleges and universities today, and as such, merits special consideration here.

The Master's Thesis

For the average graduate student, the Master's thesis is an intense and meaningful research effort which culminates his academic progress toward the Master of Arts or Master of Science degree. Through this scholarly work, he not only succeds in demonstrating his own mastery of research tools and methods, but he joins the intellectual ranks of scholars reaching back to the medieval universities of Europe. To be sure, the idea of the thesis has changed since its beginning in those early universities. At that time, the thesis was a proposition supported by the fledgling scholar. His academic ability was demonstrated by his public defense of the proposition, essentially a display of his speaking and debating ability, although knowledge of the subject was also necessary. This system prevailed in European and American universities until fairly recent times when emphasis was shifted to the research aspect. Research is the dominating theme of the thesis today.

The Master's thesis is commonly regarded as an intensive research work which allows the writer to demonstrate his ability to conceive a problem and proceed to its solution. It embodies both the processes and the results of his investigation. It is usually different from any other research that the student may have previously undertaken, although the difference may be more a matter of degree than of kind. For most, it is the first attempt to produce a scholarly work which is published, bound, and shelved. Thus, it bears the writer's name, and for all time, the quality of his scholarship will be linked to that work. In the event that he does not publish subsequent studies, the thesis will remain the major measure of his quality as a scholar.

In some respects, the thesis resembles the term paper required in undergraduate classes. It is an individual work which is systematically pursued and reported. The thesis is usually longer than the term paper since it represents a research effort extended over a greater period of time. However, length is not necessarily a measure of the quality of the thesis. In fact, the thesis problem may be more limited and specific than the term paper subject. The term paper often has a general topic which is actually no more than a guiding statement or "theme" around which the research focuses. The thesis has a specific problem for which a solution is sought. The emphasis of the term paper is commonly placed on demonstrating the skills of adequate library research and its proper reporting through a written paper. The thesis relies more heavily on the use of primary sources both in and out of the library. Both must be reported in a scholarly manner, but the thesis must do much more. It must show the reader the processes and reasoning, as well as the results of the research, and it must evaluate and present the evidence which leads to the conclusion.

In summary, the Master's thesis today in American education is a major criterion in measuring the student's scholarly abilities. It is a tangible example of his scholarship. With it as his evidence he stands ready to be judged by the masters in his selected field of study, masters who will determine if he qualifies to join their ranks. Through the thesis, he demonstrates that he can conceive and state a problem in geography and can proceed to locate and analyze the data that are related to his problem. His mastery of the analytic tools and techniques necessary for scholarly success in geography should be adequately evident. His conclusions, properly drawn from the thesis study, should be stated in a manner that evidences a high degree of reasoning ability and command of the scientific method of inquiry. It is worth repeating that it is the research method, especially the reasoning displayed in its application, which is of major concern to those who judge his petition to become a master scholar.

Organization of the Thesis

The thesis, when it is complete, is presented in the form of a book. Its organized form is basically that of other scholarly documents. This organization is presented in three sections: (1) the preliminaries, (2) the text, and (3) the reference material. Each section is vital to the whole, and the finished work is not complete without proper attention to all its parts. An abbreviated suggested outline of the proper thesis organization follows. Some geography departments might vary the order slightly, but such variations are usually minor.

I. The Preliminaries
   A. Title
   B. Submission statement
   C. Acknowledgments and/or preface
   D. Table of contents
   E. List of tables
   F. List of illustrations
II. The Text
   A. Introduction
   B. Substantive report
   C. Conclusion
III. The Reference Materials
   A. Bibliography
   B. Appendixes

If the writer keeps the primary objective of the thesis clearly in mind, the organization of the material becomes quite logical in its arrangement. Following the standard preliminaries, the first part of the text introduces the problem and establishes the ground rules to be followed in its solution. In the second and third parts, the data collected are analyzed, synthesized, and evaluated in order to test the hypotheses as solutions to the problem. The final part of the report is a listing of sources referred to in the study, as well as additional material considered pertinent and useful to future researchers.

THE PRELIMINARIES

The thesis begins with the title page, just as a book begins with the title. Of course, the bound copy has a cover and perhaps one or more blank pages before the title; but the first page of the thesis, counted as page one in small Roman numerals, is the title page. The exact content and form of this page varies slightly, depending upon the individual institution. Commonly, it contains not only the title and the author's name, but also the date the degree will be awarded, the school, and the degree for which the thesis is submitted as a partial requirement.

The title page is followed by a page called a "submission statement." This page contains the title and author's name again. The date of the submission page is the date of thesis approval by the faculty; usually, the date of the oral defense is the one designated to be shown here. Following the date, space is left for the thesis committee members and, perhaps, the dean of the graduate college to sign approval. This page is part of the finished thesis and states that the faculty has approved of the scholarship demonstrated in the written report. It may thus logically be assumed that the thesis represents their standard of excellence for a master scholar.

In many books, a statement of the aims and nature of the work to follow is made in an opening statement called the preface. Since the thesis includes a careful appraisal of the purpose, scope, and anticipated results of the research in an introductory chapter, it is not usual to find this material in a preface. In its place is usually found an acknowledgment page which is a short, simple, and tactful statement acknowledging those persons who contributed significantly and beyond normal expectations to the completed work.

The preliminary section of the thesis is completed with the table of contents, a list of tables, and a list of illustrations. The pages allow the reader immediately to observe the organization of the work and to locate the pages which contain the parts of the study in which he might be interested. Since the thesis does not commonly contain an index, the reader must rely heavily on the page number references in the table of contents and the lists of illustrations and tables.

THE TEXT

The heart and purpose of the thesis is the text. It is essentially composed of an introduction, the substantive report, and a summary, or concluding chapter. There is no set number of chapters for the three parts of the text. It is usual for the introduction to be presented in one or two chapters. The concluding section is usually one chapter in length. The substantive report itself may be one or more chapters, depending upon the logical division of the material into identifiable divisions. The parts of the thesis text, presented in a desirable and acceptable order, are shown below. However, no set order of presentation is required, and practice varies widely.

A. Introduction
1. Justification and need for the study
2. Precise statement of the research problem and research area
3. Precise statement of the hypotheses
4. Definition of terms
5. Limitations of the study
6. Review of the pertinent literature
7. Organization of the study
B. Substantive Report
1. Presentation of data
2. Analysis of data
3. Synthesis of data
4. Testing procedures
C. Conclusion
1. Summary of findings
2. Conclusions, development of theory, predictions, and recommendations

The Introduction

The introduction to the thesis informs the reader of its nature and establishes the framework. It describes the area of felt need, defines the problem, and states the possible solutions which will guide the course of study. The justification should modestly present the importance of the study to the discipline of geography. It illuminates the gaps in the knowledge, gaps which are to be filled by the thesis research.

Since geography is essentially concerned with areal relations, the portion of the earth's surface to be studied should be delimited and the scale of the study given.

In proposing the problem for the study, it is necessary that the writer define any terms which might otherwise be vague to the reader. The selection of the terms to define is left to the judgment of the writer; and certainly, the writer may expect a certain level of competence from the reader of a thesis. However, any word or concept which is not in normal usage among the anticipated readers, or any word that is used in a context different from the standard, common usage should be carefully defined.

Sometimes it becomes advisable for the writer to clarify limitations of his proposed study. Especially is this true if the title or the problem seems to suggest results more extensive than anticipated. It is not appropriate for the thesis writer to imply, intentionally or not, that he will explain something which the final study does not consider. This would be soliciting readers under false pretenses and, perhaps, obtaining tentative committee approval for the research problem as a result of this misunderstanding.

No thesis in geography should be written without knowledge of previous research related to the problem. Investigations which have already been conducted are presented both for the protection of the writer and for the edification of the reader. Since a great part of the value of scientific research is in its aid to future scholars, the history and significant trends of prior research, along with the outstanding contributions to the subject, become most important. For the same reason, the major gaps in the literature should be indicated in order to conserve the time of future researchers.

Since the review of the literature is so important, it is not uncommon for the Master's thesis to contain a separate chapter on the subject. This represents a thorough review of the geographic works related to the thesis problem and topic, as contrasted to the preliminary survey made for the work plan.

In many theses, a chapter on the organization of the study is included. Such a chapter is considered essential in some disciplines, where method and technique are emphasized more than they are in geography. If this information is not the theme of a separate chapter, it should be given prominent consideration as a chapter subdivision. The writer should explain the methods and techniques which are employed to organize and test the material. He should also outline the process followed in the study in order that the reader may understand exactly why the various steps were taken and in what order he will encounter them.

The Substantive Report

After completing the introductory portion of the thesis, the writer is ready to report the essential findings of the study. In terms of the scientific method, this includes the collection, classification, analysis, synthesis, and testing of the data so that the conclusions of the study may be made evident. This section may be presented in one or several chapters, depending upon the length of the study, the techniques employed, the areas of research, or other factors. If more than one chapter is presented, the relationship of each to the whole must be clear. It is appropriate in this respect to begin an individual chapter with a statement of the purpose of the chapter and to end with a statement of the chapter's contribution to the thesis.

It is apparent that the material embodied in this section of the thesis is the essence of the entire report. In a sense, it *is* the thesis. The other sections of the research first introduce and then conclude the report; but they are peripheral and subordinate to the section of the text referred to here as the substantive report. It is here that the research ability of the potential master scholar is displayed. This is the place to bring into focus the skills of data collecting, the detective instinct, the objectivity, the techniques, and the logical processes of reasoning which have been honed and developed in the preparation of the report.

The Conclusion

The final chapter of the report is a summary of the preceding chapters. It is also appropriate to include suggestions for further studies and to restate, in a more succinct manner, conclusions evident from the previous chapters. However, with the exceptions listed below, no new material should be introduced at this point. The final chapter should be reserved for the concluding and summarizing remarks which artistically and effectively conclude the study.

Since the summary is a recapitulation of the significant ideas and findings of the thesis, it is an important chapter. The substance recounted should be stated in the words and thoughts of the writer. It is imperative that he have mastery of the research material if he is to be capable of wording the summary well. It is this chapter that a reader might peruse to discover the crux of the study and its significant findings without referring to other sections of the report which present the data and rationale upon which the results were based.

There is one type of "new" material which might be introduced into the final chapter of the thesis. This is a statement of the writer's opinions and conclusions, *based* on his research study. The custom in this regard varies with geography departments, and the thesis writer should confer with his faculty advisor on this matter. Less questionable is the custom of suggesting future research which might be fruitfully undertaken in related areas of geography.

THE REFERENCE MATERIALS

The bibliography is an important part of the thesis or of any research report. It is the formal collection of sources related to the research problem, and it should be compiled with care. A "complete" bibliography is one which lists all works the writer has read for the study, regardless of their quality or actual contribution to the research. A "selected" bibliography is usually considered preferable since it is more helpful to future researchers. It lists all works cited or reviewed in the text and all literature which actually contributed to the solution of the problem.

The mechanics of a bibliography for geographic reports do not differ from bibliographies in other fields. Thus, only a few guidelines are reviewed here. The student is advised to consult a complete manual on thesis writing; his department quite possibly will recommend a specific one. The following are suggestions concerning the bibliography:

1. If more than twenty-five sources are listed, it is advisable to group them under headings such as "books," "periodicals," "newspapers," "government publications," "theses and dissertations," and whatever groupings seem appropriate.

2. The titles in each bibliographic group are alphabetized according to the author's last name or, in the absence of the author's name, the first word of the article.

3. Each entry is single-spaced, with the first line beginning at the left margin, and the following lines indented four spaces.

4. A double space is placed between each entry.

5. The form of all entries must be consistent.

6. Commonly, each bibliographic entry contains the author's name (last name first,) the title of the article and/or book, the place of publication, the publisher, and the date of publication.

An appendix is not required in every thesis. When it is included, it follows the bibliography and is preceded by a single sheet entitled APPENDIX. Proper contents of the appendix include tables, charts, graphs, diagrams, documents, or similar presentations of data relating to the research problem but not conveniently a part of the text. Any pertinent material which is not readily available to the reader may also be included in the appendix. If a large number of entries is included, it is wise to classify and group them as Appendix A, Appendix B, and so forth, with a title for each appendix.

Evaluating the Thesis

The final evaluation is a necessary but, in some ways, an anticlimactic part of the thesis process. If the thesis has been properly conceived, researched, and reported, the writer need have no worry concerning its acceptance. In fact, the oral defense becomes for the Master's candidate an opportunity to discuss with others a subject on which he has become something of an expert. The thesis is judged on the same points that were itemized in the early stages of planning when the operational plan was designed. There is no general agreement as to the emphasis to be placed on the various items of the thesis in the evaluation. There is common agreement, however, as to the minimum essentials necessary to meet the quality of the standard thesis. The prerequisites should be checked by the writer carefully before submitting the finished work; for if any of the items are not present, the thesis would probably be unacceptable to the judging and evaluating committee.

The following list of items includes the essential requisites for the thesis. It was compiled by John C. Almack in 1930 and has served nearly a half century of thesis writers as a guide.[2] Although it is called a checking schedule for the thesis, it is not a measuring tool and no values are assigned to the various items.

I.   The thesis is a contribution
  A.   To knowledge, truth, or
  B.   To technique or method or
  C.   Knowledge made available not before available

II.   The thesis is original
  A.   In data or principle or
  B.   In technique or method

2. John C. Almack, *Research and Thesis Writing* (Boston: Houghton Mifflin Co., 1930), p. 288. Used by permission of the publisher.

III. The method is scientific
  A. Normative or
  B. Experimental or
  C. Historical
IV. The results are scientific
  A. A norm or
  B. A law or
  C. History or
  D. New data brought under acceptable principle
 V. Requirements of the research process have been met when
  A. There is a problem
  B. There is an hypothesis
  C. The tests of it have been thorough
  D. The source is valid
  E. The data is reliable
VI. The mechanics are correct when
  A. The literature has been reviewed
  B. The introduction is complete
  C. There is a table of contents
  D. There are no typographical and grammatical errors
  E. The charts and tables are in proper form
  F. The conclusion is complete
  G. The bibliography is adequate and in proper form
  H. The form, arrangement, and binding are correct

The Abstract

An abstract of a research report has many uses. It is a capsule account which informs readers about the research, how it was done, and what was accomplished. Abstracts are published in a special compilation for theses, and often they must be submitted before a paper is accepted for presentation at scholarly meetings.

The abstract is composed of three parts which correspond to the three sections of the research text: (1) the statement of problem, (2) the exposition of methods and procedures employed in the study, and (3) a condensed summary of the findings. It is not uncommon to find abstracts composed of three paragraphs which condense the three sections into a very few sentences. The length varies, depending upon the research report and its problem, procedures, and findings.

The first part of the abstract describes the problem and the study area. For example, "The objectives of this study are (1) to describe the pattern of Tennessee voting behavior, (2) to measure the pattern in a manner capable of being duplicated, and (3) to explain the spatial variations of the pattern. The observation units are the ninety-five counties of Tennessee presidential records from 1800 to 1970."

The second section of the abstract defines the manner in which the research has been conducted and the methods, techniques, and tools employed: "Hypotheses are formulated concerning the association between the voting patterns (Y) and three independent variables (X) hypothesized as related to the pattern— (1) tradition, (2) race, and (3) income. The tests used to determine the distance, degree, and direction of the association among the variables are simple, partial and multiple correlation, and regression analysis."

The final section of the abstract informs the reader of the conclusions and uses of the study. For example, "A highly significant (.01) coefficient of correlation was found to exist between the voting pattern and the independent variables, R = .968. Among the variables, tradition (X_1) explained .681 of the "Y" distribution while race (X_2) and income (X_3) explained .421 and .201, respectively. The techniques employed should be applicable to other studies of voting patterns in other states, and results obtained should be directly comparable with those of future studies."

Abstracts may be longer than the above example, and the phraseology will vary in each case.

The Review

The review is a critical appraisal of a book or article and is usually written by a specialist on the topic or area treated in the report. It is written for those who desire to know the content of a work before purchasing or reading it. The student researcher should be familiar with reviews, because *his* research reports may be reviewed, or because he may be asked to review an article in the future. The reviewer should be objective and should not allow personal preference or bias to influence his judgment. It is his right, in fact, his obligation, to state his opinions; but they must be informed opinions and not merely criticism.

Whether the review is long or short, certain essentials should be included. They are: (1) bibliographical information, including the price, (2) the purpose of the work and for whom it is intended, (3) how it is organized, (4) what it says, (5) critical comments, (6) overall value and summary, and (7) the credentials of the author. A hypothetical example utilizing the proper form follows. The numbers are inserted as an aid in identifying the separate parts of the review:

(1) *Population Geography of New England*, John Doe. Dubuque, Iowa: Wm. C. Brown Company Publishers, 1971, 200 pp., 49 maps, 20 tables, 40 charts, 21 diagrams, bibliography, $17.50.

(2)This is a short, encyclopedic work published as a part of the School Series in Geography for those teaching population geography and related courses. (3)Written in the traditional style of population geography, the work reflects that the author has gone to considerable lengths to obtain statistical information about population patterns in the Northeastern United States. (4)Much of the textual material is intended to explain or clarify the many charts and tables presented at a rate of approximately one per page. (5)An objection to the large amount of such material is that much of it is based on 1960 census data which are already out of date. (6)While one might desire a greater emphasis on theoretical framework and more emphasis on analysis, the publication still becomes necessary reading for those seeking a complete understanding of population patterns in New England. (7)Dr. Doe is Professor of Geography at New England University and has previously authored numerous articles on population geography and demography.

appendixes

Appendix A — Selected References

The following list of references and source materials is not intended to be complete. A comprehensive listing would be so large as to be impractical for a book of this nature. The listing does, however, include many of the more widely used references in geographic research.

I. GENERAL REFERENCES—THE NATURE AND SCOPE OF MODERN GEOGRAPHY

ACKERMAN, EDWARD A. *Geography as a Fundamental Research Discipline.* Department of Geography Research Paper No. 53. Chicago: University of Chicago Press, 1958.

—— ——., ed. *The Science of Geography.* Washington, D. C.: National Academy of Science-National Research Council, 1965.

BROEK, JAN O. M. *Compass of Geography.* Columbus, Ohio: Charles E. Merrill Publishing Co., 1966.

COHEN, SAUL B., ed. *Problems and Trends in American Geography.* New York: Basic Books, 1967.

COOKE, R. U., and JOHNSON, J. H. *Trends in Geography: An Introductory Survey.* London: Pergamon Press, 1969.

FREEMAN, T. W. *A Hundred Years of Geography.* Chicago: Aldine Publishing Co., 1961.

FUSON, ROBERT H. *A Geography of Geography.* Dubuque, Iowa: Wm. C. Brown Company Publishers, 1969.

HAGGETT, PETER. *Locational Analysis in Geography.* New York: St. Martin's Press, 1965.

HARTSHORNE, RICHARD. *Perspective on the Nature of Geography.* Chicago: Rand McNally & Co. (for the Association of American Geographers), 1959.

JAMES, P. E., and JONES, C. F., eds. *American Geography*: *Inventory and Prospect*. Syracuse, N. Y.: Syracuse University Press, 1954.

SCHMEIDER, A. A.; GRIFFIN, P. F.; CHATHAM, R. L.; and NATOLI, S. J. *A Directory of Basic Geography*. Boston: Allyn & Bacon, 1970.

TAAFFE, EDWARD J., ed. *Geography*. Englewood Cliffs, N. J.: Prentice-Hall, 1970.

TAYLOR, GRIFFITH, ed. *Geography in the Twentieth-Century*. New York: Philosophical Library, 1957.

WOOLDRIDGE, S. W., and EAST, W. G. *The Spirit and Purpose of Geography*. New York: G. P. Putnam's Sons, 1967.

WRIGHT, JOHN K. *Human Nature in Geography*. Cambridge, Mass.: Harvard University Press, 1966.

II. SPECIFIC REFERENCES—RESEARCH METHODS

A. Field and Cartographic Methods

GREENHOOD, DAVID. *Mapping*. Phoenix Science Series. Chicago: University of Chicago Press, 1964.

GUNN, ANGUS M. *Techniques in Field Geography*. Toronto, Ont.: Copp Clark Pub. Co., 1962.

HART, JOHN FRASER, ed. *Field Training in Geography*. Commission on College Geography, Technical Paper No. 1, Washington, D. C.: Association of American Geographers, 1968.

JONES, C. F., and PICO, R., eds. *Symposium on the Geography of Puerto Rico*. Rio Piedras, Puerto Rico: University of Puerto Rico Press, 1955.

LOW, JULIAN W. *Plane Table Mapping*. New York: Harper & Row, Publishers, 1952.

Manual of Photographic Interpretation. Washington, D. C.: American Society of Photogrammetry, 1961.

MONKHOUSE, F. J., and WILKINSON, H. R. *Maps and Diagrams*: *Their Compilation and Construction*. Toronto, Ont.: Methuen Publications, 1952.

PLATT, ROBERT S. *Field Study in American Geography*. Department of Geography Research Paper No. 61. Chicago: University of Chicago Press, 1959.

ROBINSON, ARTHUR H. *The Look of Maps*: *An Examination of Cartographic Design*. Madison, Wis.: University of Wisconsin Press, 1952.

Rural Land Classification Program of Puerto Rico, The. Northwestern University Studies in Geography, No. 1. Evanston, Ill.: Northwestern University Press, 1952.

WHEELER, K. S., and HARDING, M. *Geographical Field Work*. London: Anthony Blond, 1965.

B. Statistical Methods

BERRY, B. J. L., and MARBLE, D. F., eds. *Spatial Analysis*: *A Reader in Statistical Geography*. Englewood Cliffs, N. J.: Prentice-Hall, 1968.

CHORLEY, R. J., and HAGGETT, P., eds. *Models in Geography.*
Toronto, Ont.: Methuen Publications, 1967.

COLE, J. P., and KING, C. A. M. *Quantitative Geography.* New
York: John Wiley & Sons, 1968.

KING, LESLIE J. *Statistical Analysis in Geography.* Englewood
Cliffs, N. J.: Prentice-Hall, 1969.

SMITH, R. H. T.; TAAFFE, E. J., and KING, L. J. *Readings in
Economic Geography: The Location of Economic Activity.*
Chicago: Rand McNally & Co., 1968.

YEATES, MAURICE H. *An Introduction to Quantitative Analysis
in Economic Geography.* New York: McGraw-Hill Book
Co., 1968.

C. Geographical Writing

DURRENBERGER, ROBERT W. *Geographical Research and Writing.* New York: Thomas Y. Crowell Co., 1970.

III. SELECTED GEOGRAPHICAL PERIODICALS AND SERIALS

A. United States and Canada

Annals of the Association of American Geographers. Association of American Geographers, 1710 Sixteenth Street N. W.,
Washington, D. C. 20036.

California Geographer, The. Journal of the California Council
for Geography Teachers, Department of Geography, California State College, Long Beach, Calif. 90801.

Commission on College Geography. (Regular Series; Resource Papers; Technical Papers). Association of American
Geographers, Arizona State University, Tempe, Ariz. 85281.

Department of Geography Research Papers. Department of
Geography, University of Chicago, 1101 East 58th Street,
Chicago, Ill. 60637.

Discussion Paper Series. Department of Geography, University of Iowa, Iowa City, Iowa 52240.

Discussion Papers. Department of Geography, The Ohio State
University, 1775 South College Road, Columbus, Ohio 43210.

East Lakes Geographer, The. Journal of the East Lakes
Division, Association of American Geographers, Depart-
ment of Geography, 1775 South College Road, Ohio State
University, Columbus, Ohio 43210.

Economic Geography. Department of Geography, Clark University, Worcester, Mass. 01610.

Focus. The American Geographical Society of New York,
Broadway at 156th Street, New York, N. Y. 10032.

Geographical Bulletin, The. Geographical Branch, Department of Mines and Technical Surveys, Ottawa, Canada.

Geographe Canadien (Canadian Geographer). Canadian Association of Geographers, Morrice Hall, McGill University,
Montreal 2, P. Q., Canada.

Geographical Review, The. The American Geographical Society of New York, Broadway at 156th Street, New York,
N. Y. 10032.

Journal of Geography, The. National Council for Geographic Education, 111 West Washington Street, Chicago, Ill. 60602.

Monograph Series of the Association of American Geographers. Rand McNally & Co., Chicago, Ill. 60680 (for the Association of American Geographers).

National Geographic Magazine. The National Geographic Society, Seventeenth and M Streets, N. W., Washington, D. C. 20036.

Professional Geographer, The. Forum and Journal of the Association of American Geographers, 1710 Sixteenth Street, N. W., Washington, D. C. 20036.

Southeastern Geographer, The. Journal of the Southeastern Division, Association of American Geographers, Department of Geography, University of North Carolina, Chapel Hill, N. C. 27514.

Soviet Geography: Review and Translation. The American Geographical Society of New York, Broadway at 156th Street, New York, N. Y. 10032.

Studies in Geography. Department of Geography, Northwestern University, Evanston, Ill. 60201.

B. Others

Australian Geographer, The. The Geographical Society of New South Wales, Department of Geography, University College, Newcastle, N. S. W., Australia.

Australian Geographical Studies. Journal of the Institute of Australian Geographers, Department of Geography, University of Melbourne, Parkville, N. 2, Victoria, Australia.

Geographical Journal, The. The Royal Geographical Society, London, S. W. 7, England.

Geographical Magazine, The. Geographical Magazine, Ltd., Friars Bridge House, Queen Victoria Street, London, E. C. 4, England.

Geography. George Philip & Son, 32 Fleet Street, London, E. C. 4, England.

New Zealand Geographer. The New Zealand Geographical Society, Department of Geography, University of Canterbury, Christchurch, New Zealand.

Publications of the Institute of British Geographers. The Institute of British Geographers, Department of Geography, Cambridge University, Cambridge, England.

Scottish Geographical Magazine, The. The Royal Scottish Geographical Society, Synod Hall, Castle Terrace, Edinburgh, Scotland.

> *NOTE:* The above list of periodicals and serials is selective and includes only American, Canadian, Australian, United Kingdom and New Zealand publications. For a comprehensive listing of geographical periodicals in English, including those of other countries, the research student is advised to refer to:

Harris, Chauncy D. *Annotated World List of Selected Current Geographical Serials in English*, 2d ed., Department of Geography, University of Chicago, 1964.

IV. BIBLIOGRAPHIC REFERENCES

Basic Geographical Library, A: A Selected and Annotated Book List for American Colleges. Commission on College Geography, Publication No. 2, Association of American Geographers, Arizona State University, Tempe, Ariz., 1966.

Current Geographical Publications. The American Geographical Society of New York, Broadway at 156th Street, New York, N. Y. Monthly, except July and August.

DURRENBERGER, ROBERT W. *Environment of Man: A Bibliography.* National Press Books, Palo Alto, Calif., 1970.

Geographical Bibliography for American Colleges, A. Commission on College Geography, Publication No. 9, Association of American Geographers, Arizona State University, Tempe, Ariz., 1970.

HARRIS, CHAUNCY D. *Annotated World List of Selected Current Geographical Serials in English.* 2d ed., Department of Geography, University of Chicago, Chicago, Ill., 1964.

HORNSTEIN, HUGH A. *A Bibliography of Paperback Books Relating to Geography.* NCGE Bibliography Series, National Council for Geographic Education, Chicago, Ill., 1970.

Inexpensive Books in Physical Geography. Mimeographed. Compiled by Harold A. Winters, Panel on Physical Geography, Commission on College Geography, Arizona State University, Tempe, Ariz., 1970.

VINGE, C. L. and VINGE, A. G. *U.S. Government Publications for Research and Teaching in Geography and Related Social and Natural Sciences.* Totowa, N. J.: Littlefield, Adams & Co., 1967.

V. STYLE MANUALS

CAMPBELL, WILLIAM G. *Form and Style in Thesis Writing.* 3d ed. Boston: Houghton Mifflin Co., 1969.

"Editorial Policy Statement." *Annals of the Association of American Geographers,* vol. 60, March, 1970, pp. 194-207. Washington, D. C.: Association of American Geographers, 1970.

MLA Style Sheet, The. Rev. ed. Compiled by William Riley Parker. New York: The Modern Language Association of America, 1968.

TURABIAN, Kate L. *A Manual for Writers of Term Papers, Theses, and Dissertations.* 3d ed. Chicago: University of Chicago Press, 1967.

Appendix B — Generalized Classification of the Physical Characteristics of the Land

In the event that detailed data concerning the physical nature of the land is desired, the following classification system may be used for multifeatured or fractional code mapping on aerial photographs or other types of base maps. The student researcher may wish to modify this classification to "fit" his research area better.

First Symbol — Slope
1. 0° to 3° (level to gently sloping).
2. 4° to 8° (undulating to rolling).
3. 9° to 14° (rolling to hilly).
4. 15° to 20° (hilly to steep slopes).
5. 21° to 30° (steep slopes).
6. 31° to more (very steep slopes).

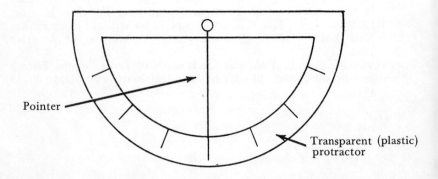

Note: Slopes may be determined and mapped by using a large protractor with a free swinging metal pointer (straighten paper clip, hat pin, etc.).

Second Symbol — Drainage

1. *Excessively drained land.* Land which, in terms of drainage, does not retain sufficient moisture for agricultural purposes.
2. *Adequately drained land.* Land which, in terms of drainage, does not normally present problems of cultivation or agricultural production, but which under heavy rainfall

   conditions, may necessitate delays in farm operations and may result in reduced crop yields.
3. *Poorly drained land.* Land which, in terms of drainage, presents problems of cultivation and agricultural production, even during conditions of normal rainfall.
4. *Very poorly drained land.* Land which, in terms of drainage, cannot be used for agricultural purposes. Virtually swamp land.

Third Symbol — Erosion

1. *Little or no observable sheet erosion.* Average removal less than 25 percent of the topsoil; if the original topsoil was about 16 inches, the present topsoil should be 12 or more inches.
2. *Moderate sheet erosion.* Average removal of from 25 to 75 percent of the topsoil; that is, if the original topsoil was about 16 inches, the present topsoil is between 4 and 12 inches.
3. *Incipient gullying.* Occasional gullies; in general, more than 100 feet apart; shallow, so that they can be crossed by tillage implements but gullies would not be obliterated by normal tillage.
4. *Damaging gully erosion.* Gullies occurring less than 100 feet apart, but not including more than 50 percent of the total mapping unit; at least one major gully not crossable with tillage implements.
5. *Excessive gully erosion.* Conditions in excess of those in #4 above. Virtually precludes cultivation.

Fourth Symbol — Surface Stoniness and Rock Exposure

1. *Free from stones.* Surface entirely free from stones, or sufficiently free from stones so that there are no difficulties in cultivating the land.
2. *Moderately stony.* Small-sized or large-sized stones at the surface in sufficient quantity to impede cultivation or render it difficult.
3. *Very stony.* Sufficient quantity or small- or large-sized stones to virtually preclude cultivation.
4. *Bedrock exposure.* Bedrock exposed, or is within plow depth of the surface and covers sufficient space to hinder severely or preclude cultivation.

Fifth Symbol — Soil Type

1. *Bottomland soils.* Clay to silt topsoil, subject to flooding.
2. *Bottomland soils.* Sand to gravel topsoil, subject to flooding.
3. *Terrace soils.* Clay to silt topsoil, normally not subject to flooding.
4. *Terrace soils.* Sand to gravel topsoil, normally not subject to flooding.
5. *Upland soils.* Clay to silt topsoil.
6. *Upland soils.* Sand to gravel topsoil.

    Note: The classification of soils above is highly generalized. The researcher is strongly advised to obtain pertinent county soil surveys developed by the Soil Conservation Service, if they are available, and modify his classification accordingly.

Appendix C — Classification of Land Use

This classification, presented here in abbreviated form, is designed to enable the researcher to collect and compile land use data by using either field mapping or aerial photo interpretation techniques. It is a stratified classification, and data may be collected in varying degrees of detail compatible to the scale of mapping. The researcher may add, or otherwise modify within the existing format, to suit his particular needs without difficulty.

First Symbol — Major Land Use

| R. RURAL LAND | | | | U. URBAN LAND |

Second Symbol — General Land Use

R. RURAL LAND USE

| 1. CROPPED LAND | 2. PASTURE LAND | 3. FOREST, GRASSLAND, SHRUB | 4. IDLE LAND | 5. MISCELLANEOUS |

Third Symbol — Specific Land Use[1]

1. CROPPED LAND[4]	2. PASTURE LAND[2]	3. FOREST, GRASSLAND, SHRUB[3]	4. IDLE LAND	5. MISCELLANEOUS
a. alfalfa	a. rotation pasture	a. spruce-fir	a. wasteland	a. airport
b. barley	b. permanent, nonwooded	b. yellow pine-Douglas fir	b. abandoned farmland	b. school
c. cotton	c. woodland pasture	c. piñon pine-juniper	c. held for non-agricultural development	c. cemetery
d. dates	d. brushland pasture	d. sagebrush (northern desert shrub)		d. park
e. corn		e. creosote (southern desert shrub)		e. mine or quarry
f. fallow		f. greasewood (salt desert shrub)		f. industry
		g. shortgrass (plains grassland)		g. golf course
etc.		h. mesquite grass (desert grassland)	etc.	etc.

Fourth Symbol — Quality

1. CROPPED LAND[4]	2. PASTURE LAND[4]	3. FOREST, GRASSLAND, SHRUB	4. IDLE LAND	5. MISCELLANEOUS
a. good quality	a. good quality	a. merchantable	(no fourth symbol)	
b. moderate	b. moderate	b. potentially merchantable		
c. poor quality	c. poor quality	c. scrub — not merchantable		

1. The RURAL LAND USE mapping key is designed for mapping large areas, i.e., fields, woodlands, etc. For small areas such as farmsteads, buildings, roads, etc., use the RURAL CULTURAL FEATURES KEY.
2. Pasture land may be defined as follows: *rotation pasture*—presently being used as pasture but will be used as a cropped land at a later date; *permanent pasture*—used as pasture year after year and does not fall into the cropped land rotation system; *woodland pasture*—at least 50 percent is covered by trees, 12 feet or higher; *brushland pasture*—at least 50 percent or more covered by brush, small trees, etc., under 12 feet in height.
3. Applies only to Southwestern United States and Northern Mexico. This category of land use must be revised to fit the particular area surveyed. Similar subcategories can be developed, however, for all sections of the country and abroad.
4. Quality may be determined subjectively by judging the general condition of the crop or pasture land. If the mapping unit is free of weeds and bare areas, and the crop is uniform in color and height, it may be judged good quality. Quality may also be determined on a measurement basis, i.e., corn yields of 100 bu. per acre—good; 70-100 bu. per acre—moderate; under 70—poor; carrying capacity of pastures, etc.

RURAL CULTURAL FEATURES KEY

First Symbol — General Functional Use

1. FARMSTEADS   2. NONFARM COUNTRY HOUSES   3. RESORT DWELLINGS   4. COMMUNAL BUILDINGS
5. COMMERCIAL BUILDINGS   6. MANUFACTURAL STRUCTURES   7. ROADS
8. IRRIGATION OR DRAINAGE DITCHES   9. AGRICULTURAL STRUCTURES

Second Symbol — Specific Functional Use

1. *FARMSTEADS*
a. beef cattle
b. citrus
c. dairy
d. poultry
e. cash grain
f. truck
etc.

2. *NONFARM HOUSES*
a. large (7 or more rooms)
b. medium (4 to 6 rooms)
c. small (less than 4 rooms)

3. *RESORT DWELLINGS*
a. year-round
b. seasonal

4. *COMMUNAL BUILDINGS*
a. school
b. church
c. club house
etc.

5. *COMMERCIAL BUILDINGS*
a. general store
b. food store
c. clothing store
d. hardware store
e. service station
f. tavern
g. drive-in theater
etc.

6. *MANUFACTURAL STRUCTURES*
a. gravel pit buildings and plant
b. quarry and buildings
c. cheese factory
d. butter factory
e. cannery
etc.

7. *ROADS*
a. paved (two lane)
b. gravel
c. improved dirt
d. unimproved
e. abandoned
f. farm lane
g. paved (four lane)

8. *IRRIGATION OR DRAINAGE DITCHES*
a. less than 5 ft. wide
b. 5 ft. - 10 ft. wide
c. 10 ft. - 20 ft. wide
d. 20 ft. or more

9. *AGRICULTURAL STRUCTURES*
a. loading pen
b. cotton gin
c. granary
d. cattle vat
e. creamery
etc.

Third Symbol — Age

1. since 1960   2. 1940-1960   3. 1920-1940   4. prior to 1920
(or other criteria suitable to the specific area)

URBAN LAND USE

Second Symbol — General Land Use

1. *RESIDENTIAL*   2. *COMMERCIAL*   3. *INDUSTRIAL*   4. *GOVERNMENTAL AND PUBLIC UTILITIES*

5. *INSTITUTIONAL*   6. *VACANT*

Third Symbol — Specific Land Use

1. *RESIDENTIAL*
   a. single family
   b. two family
   c. multifamily or apartments

2. *COMMERCIAL*
   a. appliance store
   b. bank, savings and loan
   c. cafeteria, restaurant
   d. department store
   e. drug store
   f. furniture store
   g. grocery
   h. hardware
   i. tavern, bar
   j. jewelry, gifts
   k. beauty salon
   l. men's clothing
   m. bakery

   etc.

3. *INDUSTRIAL*
   a. food and kindred products
   b. tobacco manufacturers
   c. textile mill products
   d. apparel and clothing
   e. lumber and wood products (except furniture)
   f. furniture and fixtures
   g. paper and allied products
   h. printing, publishing, and allied industries
   i. chemicals and allied products
   j. products of petroleum and coal

   etc.

4. *GOVERNMENTAL AND PUBLIC UTILITIES*
   a. court house
   b. fire station
   c. police station
   d. powerhouse or substation
   e. parks or recreation
   f. waterworks
   g. post office

   etc.

5. *INSTITUTIONAL*
   a. school
   b. church
   c. hospital

   etc.

Fourth Symbol — General Condition

1. well kept up   2. moderate condition   3. deteriorating

(Objective evaluation procedure may be defined compatible to area mapped.)

Fifth Symbol — Age

1. since 1960   2. 1940-1960   3. 1920-1940   4. prior to 1920

(or other criteria suitable to the specific area)

Appendix D — Sample Interview for Geographic Field Study

If the researcher is planning to collect data by using interviewing techniques, the sample interview sheet for agricultural studies will be of value. If the information is to be obtained indirectly by mail or other mechanisms, the questionnaire must be revised (see page 48). Similar interview sheets may be constructed for field studies in other types of geographic studies.

FARM DATA SHEET

1. *General*
   a) Name of operator _____
   b) Status (owner, manager, tenant, etc.) _____
   c) How long has he been farming? _____
   d) When did he acquire present farm? _____
   e) Original home area _____
   f) Does he work other farms? _____
   g) Is he employed in work other than farming?_____
      If so, where? _____
      Does he rent, own, sharecrop, etc.?_____
      If so, what percentage of his total income is derived from outside sources? _____

   _____
   _____

2. *Farmstead Population*

| | Family | | | Hired Help | |
Men	Women	Children (under 18)	Men	Women	Children (under 18)
_____	_____	_____	_____	_____	_____

   Number full-time _____
   Number seasonal _____

3. *Farm, Specific*
   a) Size of farm (acres) _____
   b) Type of farm (dairy, citrus, etc.) _____
   c) Acreage of tilled land (average yield per acre in parenthesis)
   Corn (____) Citrus (____) Cotton (____) Potatoes (____)
   Hay (____) Barley (____) etc. (____) etc. (____) etc. (____)
   d) Acreage of pasture _____

Rotation	Permanent Nonwooded	Wooded	Brush
_____	_____	_____	_____

   e) Other acreage _____

Timber (4 in. plus)	Brush	Wáste	Fallow	Idle land (soil bank)	Idle land (other)
_____	_____	_____	_____	_____	_____

   f) Livestock

	Dairy cattle	Beef cattle	Swine	Sheep, Poultry, etc.
Breed	_____	Breed _____	_____	_____
Milkers	_____	Raise own? _____	_____	_____
Heifers	_____	If brought in, from		
Calves	_____	where? _____	_____	_____
Bulls	_____			
Total number				
	_____	_____	_____	_____
How long has dairying been carried on?	___	_____	_____	_____
What was produced before?	_____	_____	_____	_____

   g) Farming methods
   Rotation system? _____
   Strip farm? _____If so, what fields?_____
   Contour plow? _____If so, what fields?_____
   Irrigation? _____If so, what fields?_____
   Type of fertilizer used_____
   How much applied _____ Frequency_____
   h) Income (principal commodities sold)

Commodity	% of Total Income	Where Sold
_____	. . . _____	. . . _____
_____	. . . _____	. . . _____
_____	. . . _____	. . . _____
_____	. . . _____	. . . _____

4. *Trading and Social Centers*
   School _____   Food and drugs _____
   Church _____   Hardware _____
   Movies _____   Clothing _____
   Auto service _____   Farm equipment _____
   Banking center _____   Other _____

   _____

   _____

5. *Crop hazards and general problems of the area:*

6. *Other pertinent data:*

Appendix E — Sample Checklist for Geographic Field Study

If the researcher is planning to make an analysis of the manufacturing geography of an area, the checklist below of pertinent locational and site factors will be of value. The checklist may be abbreviated or otherwise modified without difficulty. Similar checklists may be constructed for field studies in agricultural geography, settlement, trade areas, recreational geography, etc.

GENERAL INFORMATION

PLANT STRUCTURE

Products Manufactured—Type of finished product(s) presently being manufactured (if more than one, percent of total value of each); products manufactured in the past, near future (predicted mix of products and rationale).

Plant Organization—Location of branch plants, home office; policy-making structure; relationships to subsidiary plants, branches, and other types of manufacturing plants, etc.

Plant Size and Location—Total acreage, percentage used for buildings, parking, other uses, idle, etc. Location and relationship to transportation and adjacent land use.

LOCATIONAL FACTORS

RESOURCE BASE

Raw Materials—Location, availability, and price of pertinent agricultural crops, livestock products, forestry commodities, fish and marine life, minerals, other manufactured goods.

Energy—Location, availability, and price of pertinent energy sources such as coal, petroleum, gas, water power. Type of power service, reliability of service, adequacy of supply, rates, discounts, etc.

Water Supplies—Availability of surface and ground water; quality (mineral and bacteria content) for use as cleaning agent, ingredient in finished product, cooling processes, etc.

Climate—Annual, monthly, maximum-minimum temperature conditions; average degree days; annual rainfall fluctuations; frequency of fog; humidity conditions. General climatic conditions and relationships to living conditions, recreation, etc.

Land Forms and Space—Physiographic restrictions or controls, drainage, and general soil and bedrock characteristics.

ECONOMIC FACTORS

Markets—Location, size, and accessibility of markets.

Transportation—Frequency of service, freight rates, time in transit, terminal facilities, incidental costs of rail, truck, intercoastal or coastwise transportation, air cargo, and local services.

Capital—Availability, magnitude and mobility of capital, fluctuations, time factor, etc.; banking conditions.

Real Estate Value, Taxes, and Service Facilities—Tax rate of real estate and personal property, assessments, special taxes, license fees, exemptions, contemplated future changes in tax base, etc.; service facilities, such as sewers, garbage disposal, police and fire protection, streets and highways, hospital facilities, judiciary, etc.

Cost of Living—Rent, food, clothing, necessities, luxuries, residential rates, etc. on per capita basis.

Labor—Total employment; supply of suitable labor available; elements of labor unrest; past history of labor disturbances; prevailing wage scale; maximum, average, and minimum of hour shifts; labor turnover; characteristics of labor; efficiency of labor; seasonal variation; training facilities; housing; etc.

SOCIAL FACTORS

Type of Culture—Educational level and mores of community; desires to progress; capacity for accepting new ways, etc.

Organization—Organizational and cooperative aspects of community.

Civic Pride—Types and effectiveness of service and fraternal organizations; conditions of schools, churches, libraries, recreational facilities, newspapers, hotels, hospitals, and public buildings.

Technological Skills—Educational level, skills, and general intelligence of community population.

GOVERNMENTAL FACTORS
National Level—Taxes, tariffs, subsidies, etc.
Local Level—Taxes, existence and mobility of local capital, existence and type of zoning code, effectiveness and objectives of local and regional planning commissions.

SITE FACTORS

Topographic and Spatial Elevation Characteristics: Size and shape of plot; precise configuration of land; local drainage conditions; textural and compaction characteristics of the soil; exact nature of the bedrock; water table depth; frost depth; etc.

Service Facilities: Type and cost of sewer and water facilities; type of fire and police protection; garbage disposal conditions; easements; maintenance of access roads, etc.

Transportational Facilities: Rail, highway, and waterway facilities; proximity of piers, interchanges, sidings, terminals, etc.

Water Supplies: Precise volume and nature of potential surface and ground water supplies.

Land Value and Taxes: Township, county, city tax rate; real estate value and land evaluation trends.

Zoning Characteristics: Zoning restrictions, building codes, contemplated changes in zoning codes; future role of area in overall comprehensive plan, etc.

Proximity to Labor Pool: Commuting conditions, transit time; orientation aspects to local transit lines and residential areas.

Area Trends: Future growth characteristics of local surrounding areas; future restrictions on noise, odors, smoke, etc.

Appendix F — Quantitative Aids

The quantitative aids illustrated in this appendix are presented in order that the beginning research student may have available a few of the more common statistical tools used in geography. The aids presented here are not intended as a comprehensive listing, nor is it implied that these tools are suitable for all research projects. The researcher must select the best tool for the type of problem being considered. For additional quantitative aids used in geographic research, consult the bibliographic sources in Appendix A.

Correlation-Regression Analysis

TETRACHORIC CORRELATION

TABLE 9
Coefficients of Tetrachoric Correlations

%	r	%	r	%	r
45	.95	31	.37	17	−.49
44	.93	30	.31	16	−.55
43	.91	29	.25	15	−.60
42	.88	28	.19	14	−.65
41	.85	27	.13	13	−.69
40	.81	26	.07	12	−.73
39	.77	25	.00	11	−.77
38	.73	24	−.07	10	−.81
37	.69	23	−.13	9	−.85
36	.65	22	−.19	8	−.88
35	.60	21	−.25	7	−.91
34	.55	20	−.31	6	−.93
33	.49	19	−.37		
32	.43	18	−.43		

From *Short-cut Statistics for Teacher-made Tests* by Paul B. Diederich, © 1964, p. 34, Educational Testing Service, Princeton, New Jersey. Reprinted by permission of the publisher.

Read the coefficient of correlation as the number opposite the percentage of both variables above the median.

PEARSON PRODUCT MOMENT COEFFICIENT OF CORRELATION

This is a reliable and commonly used quantitative tool in all types of geographic research. It is included in the appendix rather than in the text so that it may receive individual consideration, and so that it may appear in conjunction with the form for computing correlation-regression analysis which follows it. This coefficient is obtained by first constructing a table of values for the variables being considered. The two variables, X and Y, are then squared and also multiplied in order to obtain the sum for each column to be used in the formuli. The procedure is as follows:

Step 1. Prepare a table of values as shown for the five ($N = 5$) observations below:

Obs.	X	Y	X^2	Y^2	XY
A	2	4	4	16	8
B	4	8	16	64	32
C	6	12	36	144	72
D	8	16	64	256	128
E	10	20	100	400	200
Sum	30	60	220	880	440

$N = 5$

Step 2. Compute a figure, called E, as follows:

$$E = \Sigma XY - \frac{(\Sigma X)\ (\Sigma Y)}{N}$$

Note: Σ represents the "sum of" the number it precedes.

In this example, E is thus computed:

$$E = 440 - \frac{(30)\ (60)}{5} \quad \text{or} \quad 440 - \frac{1800}{5} \quad \text{or} \quad 440 - 360$$

$$E = 80$$

Step 3. Compute two statistics conveniently called F and G. They are computed in the same manner. The formuli are:

$$F = \Sigma X^2 - \frac{(\Sigma X)^2}{N}$$

$$G = \Sigma Y^2 - \frac{(\Sigma Y)^2}{N}$$

F and G are computed as follows:

$$F = 220 - \frac{(30)\ (30)}{5} \quad \text{or} \quad 220 - \frac{900}{5} \quad \text{or} \quad 220\text{-}180$$

$$F = 40$$

$$G = 880 - \frac{(60)\ (60)}{5} \quad \text{or} \quad 880 - \frac{3600}{5} \quad \text{or} \quad 880\text{-}720$$

$$G = 160$$

Step 4. Determine the correlation coefficient by the formula

$$r = \frac{E}{\sqrt{(F)\ (G)}}$$

$$r = \frac{80}{\sqrt{(40)\ (160)}} \quad \text{or} \quad \frac{80}{\sqrt{6400}} \quad \text{or} \quad \frac{80}{80}$$

$$r = 1.00$$

Step 5. Square r to obtain the coefficient of determination, if desired.

$r^2 = $ coefficient of determination

$r^2 = 1^2$

$r^2 = 1$

REGRESSION ANALYSIS

The regression analysis results in a statement of how much the variation of the measurement on the vertical axis (Y) is related to variation of the measurement of the variable (X) on the horizontal axis. The formula for the regression line (Yc) is Yc = a + bX, where a is the elevation of the line at Y axis, and b is the slope of the line.

Using the same table of values and symbols as in the preceding example for the computation of r, the regression line (b) is determined thus:

$$b = \frac{E}{F} \text{ or } b = \frac{80}{40}$$

$$b = 2$$

The value of the Y intercept (a) in the regression equation is determined by the following formula:

$$a = \frac{\Sigma Y - b\Sigma X}{N} \text{ or } \frac{60 - (2)\ (30)}{5} \text{ or } \frac{60\text{-}60}{5} \text{ or } \frac{0}{5}$$

$$a = 0$$

Consequently, the Y computed by the regression analysis is Yc = 0 + 2X.

FORM FOR SIMPLE LINEAR REGRESSION AND
CORRELATION ANALYSIS (PEARSONIAN)

Supply the answer for: Source:
  1. Number (N) of observations _____   Table_____
  2. Sum (Σ) of columns   Table_____

ΣX _____
ΣY _____
ΣX^2 _____
ΣY^2 _____
ΣXY _____

3. $E =$ _____ $\Sigma XY - \dfrac{(\Sigma X)(\Sigma Y)}{N}$

4. $F =$ _____ $\Sigma X^2 - \dfrac{(\Sigma X)^2}{N}$

5. $G =$ _____ $\Sigma Y^2 - \dfrac{(\Sigma Y)^2}{N}$

6. $r =$ _____ $\dfrac{E}{\sqrt{(F)\ (G)}}$

7. $r^2 =$ _____ $(r)\ \ (r)$

8. $b =$ _____ $\dfrac{E}{F}$

9. $a =$ _____ $\dfrac{\Sigma Y - b(\Sigma X)}{N}$

10. $Yc =$ ____$\pm$__(X) $a + b$

TABLE 10

Significant Values of r, at .05 and .01, for Various Degrees of Freedom

1	.997		24	.388
	1.000			.496
2	.950		25	.381
	.990			.487
3	.878		26	.374
	.959			.478
4	.811		27	.367
	.917			.470
5	.754		28	.361
	.874			.463
6	.707		29	.355
	.834			.456
7	.666		30	.349
	.798			.449
8	.632		35	.325
	.765			.418
9	.602		40	.304
	.735			.393

10	.576	45	.288
	.708		.372
11	.553	50	.273
	.684		.354
12	.532	60	.250
	.661		.325
13	.514	70	.323
	.641		.302
14	.497	80	.217
	.623		.283
15	.482	90	.205
	.606		.267
16	.468	100	.195
	.590		.254
17	.456	125	.174
	.575		.228
18	.444	150	.159
	.561		.208
19	.433	200	.138
	.549		.181
20	.423	300	.113
	.537		.148
21	.413	400	.098
	.526		.128
22	.404	500	.088
	.515		.115
23	.396	1000	.062
	.505		.081
		x	

Source: William Hays. *Basic Statistics*. Brooks Cole, Belmonte, California, 1967. p. 114.

Locational Quotient

The locational quotient (LQ) of a place is a measure of the extent to which a subregion has a proportionate share of any particular phenomena found in a larger area. It is a ratio of a ratio and is expressed as:

$$LQ = \frac{\text{the ratio of phenomena X in a subdivision}}{\text{the ratio of the same phenomena in the larger division}}$$

For example, if a state had 25% of its total population employed, but a city within the state had 50% of its population employed, the LQ of the city would be 2, that is,

$$LQ = \frac{50\%}{25\%} \quad \text{or } 2$$

Table of Random Numbers

The table of random numbers is used to select a sample from any universe which has been or can be numbered. To select the sample, use the same number of digits from the column in the table as is in the total number of the universe. That is, if the total of the universe is 8,000, four digits, use all four numbers in the column. The starting place in making the sample may be determined at random, or it may be the beginning of any group in the table. After the first number, continue selecting all numbers from left to right, as in reading. If a number is larger than the total of the universe, ignore it and proceed to the next number until the desired sample is obtained. For example, if from a universe of 8,000, a sample of 80 is desired, and the starting point is from group 1, line 11, column 1, the sample would be as follows: 0709, 2523, skip, 6271, 2607, and so forth, until 80 sample numbers are selected. The third number in the table, 9224, was not considered since it was larger than the universe of 8,000. It was not counted as one number in the sample of 80 numbers.

TABLE 11
Table of Random Numbers
(8,000 Numbers)

First Thousand									
1-4	5-8	9-12	13-16	17-20	21-24	25-28	29-32	33-36	37-40
23 15	75 48	59 01	83 72	59 93	76 24	97 08	86 95	23 03	67 44
05 54	55 50	43 10	53 74	35 08	90 61	18 37	44 10	96 22	13 43
14 87	16 03	50 32	40 43	62 23	50 05	10 03	22 11	54 38	08 34
38 97	67 49	51 94	05 17	58 53	78 80	59 01	94 32	42 87	16 95
97 31	26 17	18 99	75 53	08 70	94 25	12 58	41 54	88 21	05 13
11 74	26 93	81 44	33 93	08 72	32 79	73 31	18 22	64 70	68 50
43 36	12 88	59 11	01 64	56 23	93 00	90 04	99 43	64 07	40 36
93 80	62 04	78 38	26 80	44 91	55 75	11 89	32 58	47 55	25 71
49 54	01 31	81 08	42 98	41 87	69 53	82 96	61 77	73 80	95 27
36 76	87 26	33 37	94 82	15 69	41 95	96 86	70 45	27 48	38 80
07 09	25 23	92 24	62 71	26 07	06 55	84 53	44 67	33 84	53 20
43 31	00 10	81 44	86 38	03 07	52 55	51 61	48 89	74 29	46 47
61 57	00 63	60 06	17 36	37 75	63 14	89 51	23 35	01 74	69 93
31 35	28 37	99 10	77 91	89 41	31 57	97 64	48 62	58 48	69 19
57 04	88 65	26 27	79 59	36 82	90 52	95 65	46 35	06 53	22 54
09 24	34 42	00 68	72 10	71 37	30 72	97 57	56 09	29 82	76 50
97 95	53 50	18 40	89 48	83 29	52 23	08 25	21 22	53 26	15 87
93 73	25 95	70 43	78 19	88 85	56 67	16 68	26 95	99 64	45 69
72 62	11 12	25 00	92 26	82 64	35 66	65 94	34 71	68 75	18 67
61 02	07 44	18 45	37 12	07 94	95 91	73 78	66 99	53 61	93 78
97 83	98 54	74 33	05 59	17 18	45 47	35 41	44 22	03 42	30 00
89 16	09 71	92 22	23 29	06 37	35 05	54 54	89 88	43 81	63 61
25 96	68 82	20 62	87 17	92 65	02 82	35 28	62 84	91 95	48 83
81 44	33 17	19 05	04 95	48 06	74 69	00 75	67 65	01 71	65 45
11 32	25 49	31 42	36 23	43 86	08 62	49 76	67 42	24 52	32 45

Second Thousand									
1-4	5-8	9-12	13-16	17-20	21-24	25-28	29-32	33-36	37-40
64 75	58 38	85 84	12 22	59 20	17 69	61 56	55 95	04 59	59 47
10 30	25 22	89 77	43 63	44 30	38 11	24 90	67 07	34 82	33 28
71 01	79 84	95 51	30 85	03 74	66 59	10 28	87 53	76 56	91 49
60 01	25 56	05 88	41 03	48 79	79 65	59 01	69 78	80 00	36 66
37 33	09 46	56 49	16 14	28 02	48 27	45 47	55 44	55 36	50 90
47 86	98 70	01 31	59 11	22 73	60 62	61 28	22 34	69 16	12 12
38 04	04 27	37 64	16 78	95 78	39 32	34 93	24 88	43 43	87 06
73 50	83 09	08 83	05 48	00 78	36 66	93 02	95 56	46 04	53 36
32 62	34 64	74 84	06 10	43 24	20 62	83 73	19 32	35 64	39 69
97 59	19 95	49 36	63 03	51 06	62 06	99 29	75 95	32 05	77 34
74 01	23 19	55 59	79 09	69 82	66 22	42 40	15 96	74 90	75 89
56 75	42 64	57 13	35 10	50 14	90 96	63 36	74 69	09 63	34 88
49 80	04 99	08 54	83 12	19 98	08 52	82 63	72 92	92 36	50 26
43 58	48 96	47 24	87 85	66 70	00 22	15 01	93 99	59 16	23 77
16 65	37 96	64 60	32 57	13 01	35 74	28 36	36 73	05 88	72 29
48 50	26 90	55 65	32 25	87 48	31 44	68 02	37 31	25 29	63 67
96 76	55 46	92 36	31 68	62 30	48 29	63 83	52 23	81 66	40 94
38 92	36 15	50 80	35 78	17 84	23 44	41 24	63 33	99 22	81 28
77 95	88 16	94 25	22 50	55 87	51 07	30 10	70 60	21 86	19 61
17 92	82 80	65 25	58 60	87 71	02 64	18 50	64 65	79 64	81 70
94 03	68 59	78 02	31 80	44 99	41 05	41 05	31 87	43 12	15 96
47 46	06 04	79 56	23 04	84 17	14 37	28 51	67 27	55 80	03 68
47 85	65 60	88 51	99 28	24 39	40 64	41 71	70 13	46 31	82 88
57 61	63 46	53 92	29 86	20 18	10 37	57 65	15 62	98 69	07 56
08 30	09 27	04 66	75 26	66 10	57 18	87 91	07 54	22 22	20 13

From *Tables of Random Sampling Numbers* by M. G. Kendall and B. B. Smith, (London: Cambridge University Press, © 1939), pp. 2-5. Reprinted by permission of the publisher.

Table 11 Continued

Third Thousand

	1–4	5–8	9–12	13–16	17–20	21–24	25–28	29–32	33–36	37–40
1	89 22	10 23	62 65	78 77	47 33	51 27	23 02	13 92	44 13	96 51
2	04 00	59 98	18 63	91 82	90 32	94 01	24 23	63 01	26 11	06 50
3	98 54	63 80	66 50	85 67	50 45	40 64	52 28	41 53	25 44	41 25
4	41 71	98 44	01 59	22 60	13 14	54 58	14 03	98 49	98 86	55 79
5	28 73	37 24	89 00	78 52	58 43	24 61	34 97	97 85	56 78	44 71
6	65 21	38 39	27 77	76 20	30 86	80 74	22 43	95 68	47 68	37 92
7	65 55	31 26	78 90	90 69	04 66	43 67	02 62	17 69	90 03	12 05
8	05 66	86 90	80 73	02 98	57 46	58 33	27 82	31 45	98 69	29 98
9	39 30	29 97	18 49	75 77	95 19	27 38	77 63	73 47	26 29	16 12
10	64 59	23 22	54 45	87 92	94 31	38 32	00 59	81 18	06 78	71 37
11	07 51	34 87	92 47	31 48	36 60	68 90	70 53	36 82	57 99	15 82
12	86 59	36 85	01 56	63 89	98 00	82 83	93 51	48 56	54 10	72 32
13	83 73	52 25	99 97	97 78	12 48	36 83	89 95	60 32	41 06	76 14
14	08 59	52 18	26 54	65 50	82 04	87 99	01 70	33 56	25 80	53 84
15	41 27	32 71	49 44	29 36	94 58	16 82	86 39	62 15	86 43	54 31
16	00 47	37 59	08 56	23 81	22 42	72 63	17 63	14 47	25 20	63 47
17	86 13	15 37	89 81	38 30	78 68	89 13	29 61	82 07	00 98	64 32
18	33 84	97 83	59 04	40 20	35 86	03 17	68 86	63 08	01 82	25 46
19	61 87	04 16	57 07	46 80	86 12	98 08	39 73	49 20	77 54	50 91
20	43 89	86 59	23 25	07 88	61 29	78 49	19 76	53 91	50 08	07 86
21	29 93	93 91	23 04	54 84	59 85	60 95	20 66	41 28	72 64	64 73
22	38 50	58 55	55 14	38 85	50 77	18 65	79 48	87 67	83 17	08 19
23	31 82	43 84	31 67	12 52	55 11	72 04	41 15	62 53	27 98	22 68
24	91 43	00 37	67 13	56 11	55 97	06 75	09 25	52 02	39 13	87 53
25	38 63	56 89	76 25	49 89	75 26	96 45	80 38	05 04	11 66	35 14

Fourth Thousand

	1–4	5–8	9–12	13–16	17–20	21–24	25–28	29–32	33–36	37–40
1	02 49	05 41	22 27	94 43	93 64	04 23	07 20	74 11	67 95	40 82
2	11 96	73 64	69 60	62 78	37 01	09 25	33 02	08 01	38 53	74 82
3	48 25	68 34	65 49	69 92	40 79	05 40	33 51	54 39	61 30	31 36
4	27 24	67 30	80 21	48 12	35 36	04 88	18 99	77 49	48 49	30 71
5	32 53	27 72	65 72	43 07	07 22	86 52	91 84	57 92	65 71	00 11
6	66 75	79 89	55 92	37 59	34 31	43 20	45 58	25 45	44 36	92 65
7	11 26	63 45	45 76	50 59	77 46	34 66	82 69	99 26	74 29	75 16
8	17 87	23 91	42 45	56 18	01 46	93 13	74 89	24 64	25 75	92 84
9	62 56	13 03	65 03	40 81	47 54	51 79	80 81	33 61	01 09	77 30
10	62 79	63 07	79 35	49 77	05 01	30 10	50 81	33 00	99 79	19 70
11	75 51	02 17	71 04	33 93	36 60	42 75	76 22	23 87	56 54	84 68
12	87 43	90 16	91 63	51 72	65 90	44 43	70 72	17 98	70 63	90 32
13	97 74	20 26	21 10	74 87	88 03	38 33	76 52	26 92	14 95	90 51
14	98 81	10 60	01 21	57 10	28 75	21 82	88 39	12 85	18 86	16 24
15	51 26	40 18	52 64	60 79	25 53	29 00	42 66	95 78	58 36	29 98
16	40 23	99 33	76 10	41 96	86 10	49 12	00 29	41 80	03 59	93 17
17	26 93	65 91	86 51	66 72	76 45	46 32	94 46	81 94	19 06	66 47
18	88 50	21 17	16 98	29 94	09 74	42 39	46 22	00 69	09 48	16 46
19	63 49	93 80	93 25	59 36	19 95	79 86	78 05	69 01	02 33	83 74
20	36 37	98 12	06 03	31 77	87 10	73 82	83 10	83 60	50 94	40 91
21	93 80	12 23	22 47	47 95	70 17	59 33	43 06	47 43	06 12	66 60
22	29 85	68 71	20 56	31 15	00 53	25 36	58 12	65 22	41 40	24 31
23	97 72	08 79	31 88	26 51	30 50	71 01	71 51	77 06	95 79	29 19
24	85 23	70 91	05 74	60 14	63 77	59 93	81 56	47 34	17 79	27 53
25	75 74	67 52	68 31	72 79	57 73	72 36	48 73	24 36	87 90	68 02

Table 11 Continued

	1–4	5–8	9–12	13–16	17–20	21–24	25–28	29–32	33–36	37–40
				Fifth Thousand						
1	29 93	50 69	71 63	17 55	25 79	10 47	88 93	79 61	42 82	13 63
2	15 11	40 71	26 51	89 07	77 87	75 51	01 31	03 42	94 24	81 11
3	03 87	04 32	25 10	58 98	76 29	22 03	99 41	24 38	12 76	50 22
4	79 39	03 91	88 40	75 64	52 69	65 95	92 06	40 14	28 42	29 60
5	30 03	50 69	15 79	19 65	44 28	64 81	95 23	14 48	72 18	15 94
6	29 03	99 98	61 28	75 97	98 02	68 53	13 91	98 38	13 72	43 73
7	78 19	60 81	08 24	10 74	97 77	09 59	94 35	69 84	82 09	49 56
8	15 84	78 54	93 91	44 29	13 51	80 13	07 37	52 21	53 91	09 86
9	36 61	46 22	48 49	19 49	72 09	92 58	79 20	53 41	02 18	00 64
10	40 54	95 48	84 91	46 54	38 62	35 54	14 44	66 88	89 47	41 80
11	40 87	80 89	97 14	28 60	99 82	90 30	87 80	07 51	58 71	66 58
12	10 22	94 92	82 41	17 33	14 68	59 45	51 87	56 08	90 80	66 60
13	15 91	87 67	87 30	62 42	59 28	44 12	42 50	88 31	13 77	16 14
14	13 40	31 87	96 49	90 99	44 04	64 97	94 14	62 18	15 59	83 35
15	66 52	39 45	96 74	90 89	02 71	10 00	99 86	48 17	64 06	89 99
16	91 66	53 64	69 68	34 31	78 70	25 97	50 46	62 21	27 25	06 20
17	67 41	58 75	15 08	20 77	37 29	73 20	15 75	93 96	91 76	96 99
18	76 52	79 69	96 23	72 43	34 48	63 39	23 23	94 60	88 79	06 17
19	19 81	54 77	89 74	34 81	71 47	10 95	43 43	55 81	19 45	44 07
20	25 59	25 35	87 76	38 47	25 75	84 34	76 89	18 05	73 95	72 22
21	55 90	24 55	39 63	64 63	16 09	95 99	98 28	87 40	66 66	66 92
22	02 47	05 83	76 79	79 42	24 82	42 42	39 61	62 47	49 11	72 64
23	18 63	05 32	63 13	31 99	76 19	35 85	91 23	50 14	63 28	86 59
24	89 67	33 82	30 16	06 39	20 07	59 50	33 84	02 76	45 03	33 33
25	62 98	66 73	64 06	59 51	74 27	84 62	31 45	65 82	86 05	73 00
				Sixth Thousand						
	1–4	5–8	9–12	13–16	17–20	21–24	25–28	29–32	33–36	37–40
1	27 50	13 05	46 34	63 85	87 60	35 55	05 67	88 15	47 00	50 92
2	02 31	57 57	62 98	41 09	66 01	69 88	92 83	35 70	76 59	02 58
3	37 43	12 83	66 39	77 33	63 26	53 99	48 65	23 06	94 29	53 04
4	83 56	65 54	19 33	35 42	92 12	37 14	70 75	18 58	98 57	12 52
5	06 81	56 27	49 32	12 42	92 42	05 96	82 94	70 25	45 49	18 16
6	39 15	03 60	15 56	73 16	48 74	50 27	43 42	58 36	73 16	39 90
7	84 45	71 93	10 27	15 83	84 20	57 42	41 28	42 06	15 90	70 47
8	82 47	05 77	06 89	47 13	92 85	60 12	32 89	25 22	42 38	87 37
9	98 04	06 70	24 21	69 02	65 42	55 33	11 95	72 35	73 23	57 26
10	18 33	49 04	14 33	48 50	15 64	58 26	14 91	46 02	72 13	48 62
11	33 92	19 93	38 27	43 40	27 72	79 74	86 57	41 83	58 71	56 99
12	48 66	74 30	44 81	06 80	29 09	50 31	69 61	24 64	28 89	99 79
13	85 85	07 54	21 50	31 80	10 19	56 65	82 52	26 58	55 12	26 34
14	08 27	08 08	35 87	96 57	33 12	01 77	52 76	09 89	71 12	17 69
15	59 61	22 14	26 09	96 75	17 94	51 08	41 91	45 94	80 48	59 92
16	17 45	77 79	31 66	36 54	92 85	65 60	53 98	63 50	11 20	96 63
17	11 26	37 08	07 71	95 95	39 75	92 48	99 78	23 33	19 56	06 67
18	48 08	13 98	16 52	41 15	73 96	32 55	03 12	38 30	88 77	17 03
19	76 27	72 22	99 61	72 15	00 25	21 54	47 79	18 41	58 50	57 66
20	98 89	22 25	72 92	53 55	07 98	66 71	53 29	61 71	56 96	41 78
21	88 69	61 63	01 67	61 88	58 79	35 65	08 45	63 38	69 86	79 47
22	12 58	13 75	80 98	01 35	91 16	18 36	90 54	99 17	68 36	85 06
23	08 86	96 36	14 09	43 85	51 20	65 18	06 40	52 17	48 10	68 97
24	33 81	05 51	32 48	60 12	32 44	08 12	89 00	98 82	79 17	97 22
25	05 15	99 28	87 15	07 08	66 92	53 81	69 42	02 27	65 33	57 69

Table 11 Continued

Seventh Thousand

	1–4	5–8	9–12	13–16	17–20	21–24	25–28	29–32	33–36	37–40
1	80 30	23 64	67 96	21 33	36 90	03 91	69 33	90 13	34 48	02 19
2	61 29	89 61	32 08	12 62	26 08	42 00	31 73	31 30	30 61	34 11
3	23 33	61 01	02 21	11 81	51 32	36 10	23 74	50 31	90 11	73 52
4	94 21	32 92	93 50	72 67	23 20	74 59	30 30	48 66	75 32	27 97
5	87 61	92 69	01 60	28 79	74 76	86 06	39 29	73 85	03 27	50 57
6	37 56	19 18	03 42	86 03	85 74	44 81	86 45	71 16	13 52	35 56
7	64 86	66 31	55 04	88 40	10 30	84 38	06 13	58 83	62 04	63 52
8	22 69	58 45	49 23	09 81	98 84	05 04	75 99	27 70	72 79	32 19
9	23 22	14 22	64 90	10 26	74 23	53 91	27 73	78 19	92 43	68 10
10	42 38	59 64	72 96	46 57	89 67	22 81	94 56	69 84	18 31	06 39
11	17 18	01 34	10 98	37 48	93 86	88 59	69 53	78 86	37 26	85 48
12	39 45	69 53	94 89	58 97	29 33	29 19	50 94	80 57	31 99	38 91
13	43 18	11 42	56 19	48 44	45 02	84 29	01 78	65 77	76 84	88 85
14	59 44	06 45	68 55	16 65	66 13	38 00	95 76	50 67	67 65	18 83
15	01 50	34 32	38 00	37 57	47 82	66 59	19 50	87 14	35 59	79 47
16	79 14	60 35	47 95	90 71	31 03	85 37	38 70	34 16	64 55	66 49
17	01 56	63 68	80 26	14 97	23 88	59 22	82 39	70 83	48 34	46 48
18	25 76	18 71	29 25	15 51	92 96	01 01	28 18	03 35	11 10	27 84
19	23 52	10 83	45 06	49 85	35 45	84 08	81 13	52 57	21 23	67 02
20	91 64	08 64	25 74	16 10	97 31	10 27	24 48	89 06	42 81	29 10
21	80 86	07 27	26 70	08 65	85 20	31 23	28 99	39 63	32 03	71 91
22	31 71	37 60	95 60	94 95	54 45	27 97	03 67	30 54	86 04	12 41
23	05 83	50 36	09 04	39 15	66 55	80 36	39 71	24 10	62 22	21 53
24	98 70	02 90	30 63	62 59	26 04	97 20	00 91	28 80	40 23	09 91
25	82 79	35 45	64 53	93 24	86 55	48 72	18 57	05 79	20 09	31 46

Eighth Thousand

	1–4	5–8	9–12	13–16	17–20	21–24	25–28	29–32	33–36	37–40
1	37 52	49 55	40 65	27 61	08 59	91 23	26 18	95 04	98 20	99 52
2	48 16	69 65	69 02	08 83	08 83	68 37	00 96	13 59	12 16	17 93
3	50 43	06 59	56 53	30 61	40 21	29 06	49 60	90 38	21 43	19 25
4	89 31	62 79	45 73	71 72	77 11	28 80	72 35	75 77	24 72	98 43
5	63 29	90 61	86 39	07 38	38 85	77 06	10 23	30 84	07 95	30 76
6	71 68	93 94	08 72	36 27	85 89	40 59	83 37	93 85	73 97	84 05
7	05 06	96 63	58 24	05 95	56 64	77 53	85 64	15 95	93 91	59 03
8	03 35	58 95	46 44	25 70	31 66	01 05	44 44	62 91	36 31	45 04
9	13 04	57 67	74 77	53 35	93 51	82 83	27 38	63 16	04 48	75 23
10	49 96	43 94	56 04	02 79	55 78	01 44	75 26	85 54	01 81	32 82
11	24 36	24 08	44 77	57 07	54 41	04 56	09 44	30 58	25 45	37 56
12	55 19	97 20	01 11	47 45	79 79	06 72	12 81	86 97	54 09	06 53
13	02 28	54 60	28 35	32 94	36 74	51 63	96 90	04 13	30 43	10 14
14	90 50	13 78	22 20	37 56	97 95	49 95	91 15	52 73	12 93	78 94
15	33 71	32 43	29 58	47 38	39 96	67 51	64 47	49 91	64 58	93 07
16	70 58	28 49	54 32	97 70	27 81	64 69	71 52	02 56	61 37	04 58
17	09 68	96 10	57 78	85 00	89 81	98 30	19 40	76 28	62 99	99 83
18	19 36	60 85	35 04	12 87	83 88	66 54	32 00	30 20	05 30	42 63
19	04 75	44 49	64 26	51 46	80 50	53 91	00 55	67 36	68 66	08 29
20	79 83	32 39	46 77	56 83	42 21	60 03	14 47	07 01	66 85	49 22
21	80 99	42 43	08 58	54 41	98 05	54 39	34 42	97 47	38 35	59 40
22	48 83	64 99	86 94	48 78	79 20	62 23	56 45	92 65	56 36	83 02
23	28 45	35 85	22 20	13 01	73 96	70 05	84 50	68 59	96 58	16 63
24	52 07	63 15	82 30	66 23	14 26	66 61	17 80	41 97	40 27	24 80
25	39 14	52 18	35 87	48 55	48 81	03 11	26 99	03 80	08 86	50 42

index